SPRINGER TRACTS IN MODERN PHYSICS

Ergebnisse
der exakten Natur-
wissenschaften

Volume 52

Springer-Verlag Berlin Heidelberg GmbH 1970

Manuscripts for publication should be addressed to:

G. Höhler, Institut für Theoretische Kernphysik der Universität, 75 Karlsruhe, Kaiserstraße 12

Proofs and all correspondence concerning papers in the process of publication should be addressed to:

E. A. Niekisch, Kernforschungsanlage Jülich, Institut für Technische Physik, 517 Jülich, Postfach 365

ISBN 978-3-662-15891-3 ISBN 978-3-540-36316-3 (eBook)
DOI 10.1007/978-3-540-36316-3

Originally published by Springer-Verlag Berlin Heidelberg New York in 1970.

Weak Interactions

Invited Papers presented at the
second international Summer School
for Theoretical Physics
University of Karlsruhe
(July 14 — August 1, 1969)

Contents

A Survey of the Weak Interactions
S. GASIOROWICZ 1

Semileptonic Decays
V. F. MÜLLER 34

Non Leptonic Decays
B. STECH 50

Current Algebra and Weak Interactions
B. RENNER 60

The Decays of the K_0—$\overline{K}_0$ System
H. G. DOSCH 79

Questions Raised by *CP*-Nonconservation
P. K. KABIR 91

Relations for Semileptonic Weak Interactions Involving Photons
W. KUMMER 113

Radiative Corrections to Weak Decays Involving Leptons
RIAZUDDIN 126

Radiative Corrections to Weak Interactions
J. ROTHLEITNER 161

Unconventional Models of Weak Interactions
G. SEGRÈ 171

Weak Interactions at Small Distances
H. PIETSCHMANN 193

Physical Symmetries in the Framework of Quantum Field Theory
J. T. ŁOPUSZAŃSKI 201

A Survey of the Weak Interactions*

S. Gasiorowicz

Contents

Introduction 1
I. The Weak Interactions before 1957 1
II. Parity Nonconservation and the Form of the Beta Interaction 5
III. Currents and their Symmetries 14
IV. Nonleptonic Decays 25
V. Concluding Remarks 30
References 31

Introduction

The study of the weak interactions is particularly fascinating, because these interactions play a dual role in the physics of the fundamental structure of matter. On one hand they manifest themselves in the interaction of leptons (electrons, muons, neutrinos) which have no strong interactions; it is thus possible to study them in their "pure" form, contaminated only by the reasonably well-understood electromagnetic interactions. On the other hand, the strongly interacting hadrons also interact weakly, and it is possible to learn something about the strong interactions with the help of the weak ones as "probes". The degree to which the weak interactions are understood, and what they have taught us about the strong interactions, is the theme of this school. The role of this survey is to show, without going into too much detail, how the present ideas about the weak interactions have evolved, and how the various topics treated in the specialized lectures are related to each other.

I. The Weak Interactions before 1957**

Although the weak interactions, specifically beta decay, were discovered at the end of the nineteenth century, no systematic theoretical work was possible before the daring hypothesis of the existence of the

* Supported in part by the U.S. Atomic Energy Commission under contract AT (11-1)-1764.

** The notation that we use is that of *S. Gasiorowicz* "Elementary Particle Physics" John Wiley Inc. New York (1966) with one exception: the normalization of the spin 1/2 states is taken to be the same as that of the spin 0 states.

neutrino, made by *Pauli** in 1930 and the discovery of the neutron by *Chadwick* in 1932. The identification of the basic reaction as

$$n \to p + e^- + \nu$$

made possible the quantum field theory formulation of beta decay by *Fermi* (1934). *Fermi* proposed to describe the interaction by the Hamiltonian

$$H_I = C \int d^3x \{ (\overline{\psi}_n(x) \gamma_\alpha \psi_p(x)) (\overline{\psi}_\nu(x) \gamma^\alpha \psi_e(x)) + (\overline{\psi}_p(x) \gamma_\alpha \psi_n(x)) (\overline{\psi}_e(x) \gamma^\alpha \psi_\nu(x)) \} . \tag{I-1}$$

Here the quantities $\psi_n(x)$, $\psi_p(x)$, ... are field operators. The operator $\psi_p(x)$, for example, annihilates a proton or creates an antiproton, and similarly for the other operators. This dual role of the operators is necessary for the relativistic invariance of the theory, and it implies that the same interaction Hamiltonian also describes the reaction

$$e^- + p \to n + \nu$$

which is observed in the capture of electrons from atomic orbits about nuclei. The Hamiltonian also describes positron decay

$$p \to n + e^+ + \nu$$

which is energetically possible in nuclei, and the reactions

$$\nu + p \to e^+ + n ,$$

$$\nu + n \to e^- + p ,$$

which can be studied in the laboratory with high energy neutrino beams.

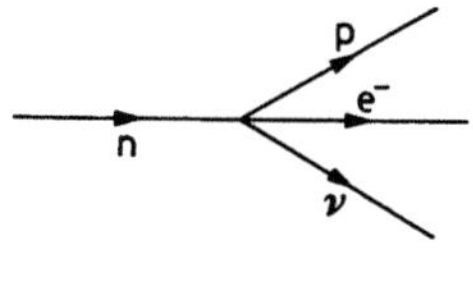

Fig. 1. Graphical representation of the Fermi interaction for the reaction $n \to p + e^- + \nu$. Related reactions are described by a reorientation of the lines with the interpretation that a change from outgoing to ingoing line requires a change from particle to antiparticle

As was already noted by *Fermi*, the Hamiltonian in (I-1) is not unique. If one assumes the same local form (which has the diagrammatic representation given in Fig. 1), does not include derivatives of field

* A very interesting historical review may be found in *C. S. Wu*, "The Neutrino", in *Theoretical Physics in the Twentieth Century*, *M. Fierz* and *V. F. Weisskopf* (Ed.), Interscience Publishers, New York (1960).

operators* and if one assumes that the interaction is invariant under space and time inversions (something which was totally taken for granted before 1957), then the most general form that one can write is

$$\begin{aligned} H_1 = \int d^3x \{ & C_s(\bar{\psi}_n(x)\, \psi_p(x))\, (\bar{\psi}_\nu(x)\, \psi_e(x)) + \text{h.c.} \\ & + C_V(\bar{\psi}_n(x)\, \gamma_\alpha \psi_p(x))\, (\bar{\psi}_\nu(x)\, \gamma^\alpha \psi_e(x)) + \text{h.c.} \\ & + \tfrac{1}{2} C_T(\bar{\psi}_n(x)\, \sigma_{\alpha\beta} \psi_p(x))\, (\bar{\psi}_\nu(x)\, \sigma^{\alpha\beta} \psi_e(x)) + \text{h.c.} \\ & + C_A(\bar{\psi}_n(x)\, \gamma_\alpha \gamma_5 \psi_p(x))\, (\bar{\psi}_\nu(x)\, \gamma^\alpha \gamma_5 \psi_e(x)) + \text{h.c.} \\ & + C_p(\bar{\psi}_n(x)\, \gamma_5 \psi_p(x))\, (\bar{\psi}_\nu(x)\, \gamma_5 \psi_e(x)) + \text{h.c.} \}\,. \end{aligned} \tag{I-2}$$

The inadequacy of (I-1) in view of the experimental situation was pointed out by *Gamow* and *Teller* (1936). Abandoning space inversion (parity) invariance allows for the addition of terms of the form

$$\begin{aligned} \int d^3x\, \{ & C'_S(\bar{\psi}_n(x)\, \psi_p(x))\, (\bar{\psi}_\nu(x)\, \gamma_5 \psi_e(x)) + \text{h.c.} \\ & + C'_V(\bar{\psi}_n(x)\, \gamma_\alpha \psi_p(x))\, (\bar{\psi}_\nu(x)\, \gamma^\alpha \gamma_5 \psi_e(x)) + \text{h.c.} \\ & + \cdots \}\,. \end{aligned} \tag{I-3}$$

Time reversal invariance requires all the C's to be real. One does not need to include terms like

$$\int d^3x \{ \hat{C}_S(\bar{\psi}_n(x)\, \psi_e(x))\, (\bar{\psi}_\nu(x)\, \psi_p(x)) + \text{h.c.} + \cdots \} \tag{I-4}$$

because of a general theorem of *Fierz* (1937) based on the completeness of the sixteen 4×4 Dirac matrices, according to which

$$\sum_i C_i(\bar{\psi}_1 \Gamma_i \psi_2)\, (\bar{\psi}_3 \Gamma^i \psi_4) = \sum_i D_i(\bar{\psi}_1 \Gamma_i \psi_4)\, (\bar{\psi}_3 \Gamma^i \psi_2) \tag{I-5}$$

where $\Gamma_i = 1, \gamma_\alpha, \frac{1}{\sqrt{2}} \sigma_{\alpha\beta}, \gamma_\alpha \gamma_5, \gamma_5$ and

$$\begin{pmatrix} D_S \\ D_V \\ D_T \\ D_A \\ D_P \end{pmatrix} = -\tfrac{1}{4} \begin{pmatrix} 1 & 4 & 6 & 4 & 1 \\ 1 & -2 & 0 & 2 & -1 \\ 1 & 0 & -2 & 0 & 1 \\ 1 & 2 & 0 & -2 & -1 \\ 1 & -4 & 6 & -4 & 1 \end{pmatrix} \begin{pmatrix} C_S \\ C_V \\ C_T \\ C_A \\ C_P \end{pmatrix}. \tag{I-6}$$

Given the interaction (I-2), it is a fairly straightforward matter to derive the experimental consequences for the allowed transitions in beta decay of nuclei, i.e. for the transitions for which the matrix elements $\langle f | \bar{\psi}_p \Gamma_i \psi_n | i \rangle$, $(\Gamma_i = \gamma_0, \gamma_k \gamma_5)$ do not vanish (the other matrix elements are very small

* Early misinterpretations of the electron spectrum measurements appeared to require a theory involving derivatives. Such a theory was discussed by *Konopinski* and *Uhlenbeck* (1941).

when the nucleon momenta are nonrelativistic). A simple calculation shows the following

$$\begin{aligned}\frac{d^2\Gamma}{dp_e\,dp_\nu} \propto \Big\{ & |C_S|^2\left(1-\frac{p_e\cdot p_\nu}{E_eE_\nu}\right)|\langle 1\rangle|^2 \\ & +|C_V|^2\left(1+\frac{p_e\cdot p_\nu}{E_eE_\nu}\right)|\langle 1\rangle|^2 \\ & +\frac{1}{4}|C_T|^2\left(1+\frac{1}{3}\frac{p_e\cdot p_\nu}{E_eE_\nu}\right)|\langle \sigma\rangle|^2 \\ & +|C_A|^2\left(1-\frac{1}{3}\frac{p_e\cdot p_\nu}{E_eE_\nu}\right)|\langle \sigma\rangle|^2 \\ & +\frac{2m_e}{E_e}C_SC_V|\langle 1\rangle|^2-\frac{m_e}{E_e}C_AC_T|\langle \sigma\rangle|^2\Big\}.\end{aligned} \tag{I-7}$$

In particular the terms proportional to C_SC_V and C_AC_T show a marked energy dependence; such an energy dependence is not observed from which one deduces that

$$\begin{aligned}C_SC_V &\approx 0,\\ C_AC_T &\approx 0.\end{aligned} \tag{I-8}$$

The choice between S and V for the transitions involving the nuclear matrix element $\langle 1\rangle = \langle f|\psi_p^+\psi_n|i\rangle$, and between T and A for the transitions involving $\langle \sigma\rangle = \langle f|\psi_p^+\sigma\psi_n|i\rangle$ can be made by means of measurements of electron-neutrino correlations. Such experiments are very difficult* and by 1957 the choice implied by (I-8) had not been made. What was known from the decay rates was that the magnitude of the largest of the C's (which have dimension M^{-2}) was

$$|C| \cong 10^{-5}\,m_p^{-2}. \tag{I-9}$$

During the late 1940's information about other weak interactions began to accumulate. The newly discovered muon was found not to interact strongly; its electromagnetic interactions were consistent with it being a heavy electron (this now seems to be true to a very high degree of accuracy) and it had interactions which led to reactions like

$$\mu^- + p \to n + \nu$$

and

$$\mu^\pm \to e^\pm + \nu + \nu\,.$$

---

* For a review see *Allen* (1959).

A very important observation made by a number of people (*Klein*, 1948; *Puppi*, 1948; *Tiomno* and *Wheeler*, 1949; *Lee*, *Rosenbluth*, and *Yang*, 1949) was that if the Hamiltonian (I-2) was also used to describe the above reactions (with the obvious substitutions) then the *C*'s appeared to be of the same order of magnitude as those for beta decay. This observation was generalized to a principle of *universality of the weak interactions*, and its formulation at the time was that in the form

$$\sum_{i=S,\ldots P} C_i(\overline{\psi}_1 \Gamma_i \psi_2)(\overline{\psi}_3 \Gamma^i \psi_4) \tag{I-10}$$

the various weak interactions could be described by taking for the pairs (1, 2) and (3, 4) pairs of fields (n, p), (ν, e) and (ν, μ). This notion proved very fruitful. It provided a mechanism from the comparison of the reaction rates for $\pi \to \mu + \nu$ and $\pi \to e + \nu$ via the mechanism

$$\pi^+ \xrightarrow{\text{strong}} (p + \overline{n}) \xrightarrow{\text{weak}} \begin{cases} \mu^+ + \nu \\ e^+ + \nu \end{cases}.$$

Although the simplest Feynman graph (Fig. 2) led to a divergent expression, the universality of the $(\mu\nu)$ and $(e\nu)$ couplings implied that information about the relative magnitudes of C_A and C_P could be obtained from the

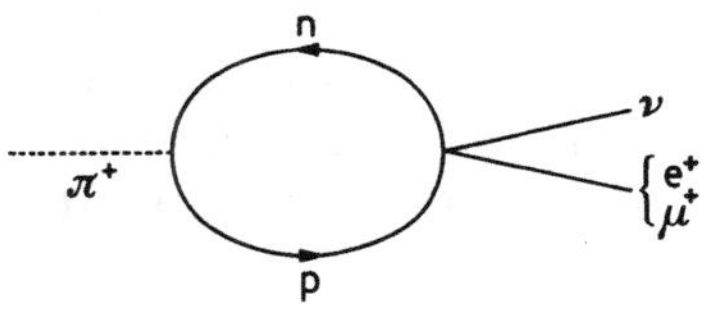

Fig. 2. Simplest Feynman graph for the decay of a charged pion

ratio $\Gamma(\pi \to e\nu)/\Gamma(\pi \to \mu\nu)$ (*Ruderman* and *Finkelstein*, 1949). Here too the information was not conclusive and the whole field seemed to be in a state of confusion (*Michel*, 1956). The confusion was soon removed by a whole reevaluation of the known experimental information and by a great deal of new experimental activity stimulated by a paper of *Lee* and *Yang* (1956) entitled "Question of Parity Conservation in Weak Interactions".

II. Parity Nonconservation and the Form of the Beta Interaction

In the early 1950's a rather cozy, if intractable, picture of a hadronic world consisting of protons, neutrons and pions had to be scrapped because of the discovery of the "strange" particles. Although their peculiar interactions were soon systematized by the introduction of the "strangeness" quantum number by *Gell-Mann* (1953) and *Nakano* and *Nishijima* (1953), very little was understood about them. They could, in principle be fitted into the "universal" picture of (I-10) by adding the pair (Λ, p) to the set of pairs, but no attempt at any theoretical work was possible before the spins and parities of the particles were determined. In trying to establish these properties for what seemed to be a single particle (same mass and lifetime within the considerable error limits) an apparent contradiction was found. In the decays

$$K^+ \to \pi^+ \pi^0$$

the final states could only have spin and parity $0^+, 1^-, 2^+, \ldots$ whereas the evidence from the decay

$$K^+ \to \pi^+ + \pi^+ + \pi^-$$

(*Dalitz*, 1953; *Fabri*, 1954; *Orear et al.*, 1956) strongly suggested the spin and parity assignment 0^-. As the masses and lifetimes of the "two" particles came closer and as the artificiality of ad-hoc explanations of the $K^+_{3\pi}$ spectrum became more evident, *Lee* and *Yang* raised the question of parity nonconservation in the weak interactions. They noted that except for this puzzle there was no evidence on this point, since all rates and correlations that had been measured until that time were *scalar observables**. To see that this is insufficient, note that parity nonconservation implies that a matrix element must contain both even and odd parity terms

$$M = M_+ + M_-$$

where $PM_\pm P^{-1} = \pm M_\pm$. Thus in a reaction rate which measures

$$|M|^2 = |M_+|^2 + |M_-|^2 + 2\,\mathrm{Re}\,M_+ M_-^*$$

the presence of *both* M_+ and M_- can only be definitely established by a measurement of $\mathrm{Re}\,M_+ M_-^*$, which is *odd* under parity and whose nonvanishing definitely proves parity nonconservation. A measurement of a decay rate, for example, involves an averaging over spins (or integration over momenta) which never detects the presence of the pseudoscalar term. *Lee* and *Yang* then suggested the measurement of some pseudoscalar observables in a weak decay. Most of their tests reduced to a measurement of quantities like $\langle \boldsymbol{S} \cdot \boldsymbol{p} \rangle$, where $\boldsymbol{S}$ is an axial vector, e.g. the polarization of a nucleus, and $\boldsymbol{p}$ is a polar vector, e.g. the momentum of the electron

* For an interesting exception see the report of *L. Grodzins* in the Proceedings of the National Academy of Sciences, Washington, **45**, 399 (1959).

emitted in the decay, as in the Co^{60} experiment which first proved parity nonconservation (*Wu*, *Ambler*, *Hayward*, *Hoppes*, and *Hudson*, 1957). The above experiment and a number of others* established that (1) parity was not conserved in the weak interactions; (2) charge conjugation was not conserved in the weak interactions; (3) to a high degree of accuracy time reversal was still a good symmetry; (4) a serviceable approximation to the weak interactions Hamiltonian for the reactions discussed above was**

$$H_{\mathrm{I}} = \frac{G}{\sqrt{2}} \int d^3x\, J^{\alpha}(x)\, J_{\alpha}^{+}(x)\,, \qquad \text{(II-1)}$$

$$J^{\alpha}(x) = \bar{\psi}_e \gamma^{\alpha}(1-\gamma_5)\,\psi_{\nu} + \bar{\psi}_{\mu}\gamma^{\alpha}(1-\gamma_5)\,\psi_{\nu} + \bar{\psi}_n\gamma^{\alpha}(1-\gamma_5)\,\psi_p + \cdots$$

which is rather close to the form originally proposed by *Fermi*!

Let us now briefly turn to the first three points before discussing the Hamiltonian (II-1).

(1) In the Co^{60} experiment the nuclei were polarized and the electrons coming from the decay

$$Co^{60} \to Ni^{60*} + e^- + \nu$$

were detected. It was found that their distribution was of the form

$$W(\theta) = 1 + \alpha \cos\theta \qquad \text{(II-2)}$$

where $\cos\theta = \hat{p}_e \cdot \langle \boldsymbol{J} \rangle / J$ and $\alpha_{\mathrm{exp}} \cong -0.4$. This observation was enough to show once for all that parity was not conserved in the decay. The magnitude of α suggested that the quantity

$$\frac{2\,\mathrm{Re}(C_T C_T'^* - C_A C_A'^*)}{|C_T|^2 + |C_T'|^2 + |C_A|^2 + |C_A'|^2} \approx 1\,.$$

This result together with the fact that the absence of the Fierz interference terms now implied that

$$\mathrm{Re}(C_A' C_T'^* - C_A C_T^*) \approx 0$$

suggested simple solutions

$$C_T = C_T'\,, \quad C_A = C_A' = 0 \qquad \text{(II-3)}$$

or

$$C_A = -C_A'\,, \quad C_T = C_T' = 0 \qquad \text{(II-4)}$$

* For a full discussion of the experiments the books by *Schopper* (1966) and *Wu* and *Moszkowski* (1966) should be consulted.

** This simple form was anticipated in theoretical speculations of *Sudarshan* and *Marshak* (1957), *Feynman* and *Gell-Mann* (1958), *Sakurai* (1958) and *Theis* (1958).

which would have the neutrino appear in the interaction Hamiltonian in the form $(1+\gamma_5)\psi_\nu$ or $(1-\gamma_5)\psi_\nu$. Since the neutrino is massless*, it is perfectly consistent to envisage a world in which only one or the other combination appears. The Dirac equation for a free particle may be written as

$$\begin{aligned} i\gamma^\mu \partial_\mu \psi_+ &= m\psi_- , \\ i\gamma^\mu \partial_\mu \psi_- &= m\psi_+ \end{aligned} \qquad \text{(II-5)}$$

where $\psi_\pm = (1 \pm \gamma_5)\psi$, and when $m=0$, these decouple. If neutrinos only interact weakly and if all interactions (not only the Gamow-Teller part measured in the Co^{60} experiment) involve the same combination, then we may restrict ourselves to a *two-component neutrino theory* (*Salam*, 1957; *Landau*, 1957; *Lee* and *Yang*, 1957) involving only one or the other combination. These combinations may be associated with the helicities of the neutrino.

Further experiments showed that the same two options were open in Fermi transitions, i.e. that possible choices were

$$\begin{aligned} C_S &= C'_S , \\ C_V &= C'_V = 0 \end{aligned} \qquad \text{(II-6)}$$

or

$$\begin{aligned} C_V &= -C'_V , \\ C_S &= C'_S = 0 \end{aligned} \qquad \text{(II-7)}$$

and that in transitions involving both Fermi and Gamow-Teller matrix elements ($J \to J (J \neq 0)$ transition) there was interference indicating that the same helicity appeared in both. Thus the choice was between

$$S, T \quad \text{and} \quad (1+\gamma_5)\psi_\nu$$

and

$$V, A \quad \text{and} \quad (1-\gamma_5)\psi_\nu .$$

The measurement of the neutrino helicity in a very ingenious experiment by *Goldhaber*, *Grodzins* and *Sunyar* (1958) showed that the second option was the one consistent with experiment.

(2) The magnitude of α also indicated that charge conjugation invariance was violated in the weak interactions. Let us see how this far-from-evident deduction follows. Consider the interaction Hamiltonian

$$\begin{aligned} H_\mathrm{I} &= \int d^3x\, \mathscr{H}_\mathrm{I}(x) , \\ \mathscr{H}_\mathrm{I}(x) &= \sum_i C_1 (\bar\psi_1 \Gamma_i \psi_2)(\bar\psi_3 \Gamma_i \psi_4) + \sum_i C_i^* (\bar\psi_2 \Gamma_i \psi_1)(\bar\psi_4 \Gamma_i \psi_3) \\ &\quad + \sum_i C'_i (\bar\psi_1 \Gamma_i \psi_2)(\bar\psi_3 \tilde\Gamma_i \psi_4) + \sum_i C'^*_i (\bar\psi_2 \Gamma_i \psi_1)(\bar\psi_4 \tilde{\tilde\Gamma}_i \psi_3) \end{aligned} \qquad \text{(II-8)}$$

* A recent limit from beta decay gives $m_\nu < 60$ eV (*Bergkvist*, 1969).

where we have now chosen the Γ_i and $\tilde{\Gamma}_i, \tilde{\tilde{\Gamma}}_i$ such that $\gamma_0 \Gamma_i^+ \gamma_0 = \Gamma_i$, and $\tilde{\Gamma}_i = \Gamma_i \gamma_5$; $\tilde{\tilde{\Gamma}}_i = -\gamma_5 \Gamma_i$.

Since under the parity transformation*

$$\begin{aligned} \psi_1 &\to \gamma_0 \psi_1 , \\ \overline{\psi}_1 &\to \overline{\psi}_1 \gamma_0 , \\ &\ldots\ldots\ldots\ldots \end{aligned} \tag{II-9}$$

an examination of the various terms shows that a change in the Hamiltonian can be described by

$$\{C_i, C_i'\} \xrightarrow{P} \eta_p \{C_i, -C_i'\} . \tag{II-10}$$

Thus an interference term between the C_i and the C_j' shows parity violation. Consider next charge conjugation. The transformation for the fields is

$$\begin{aligned} \psi_1 &\to \overline{\psi}_1 C , \\ \overline{\psi}_1 &\to -C^{-1} \psi_1 \end{aligned} \tag{II-11}$$

with

$$CC^+ = 1 ; \quad C = -C^T ; \quad C^{-1} \gamma^\mu C = -(\gamma^\mu)^T .$$

Hence

$$(\overline{\psi}_1 \Gamma \psi_2) \to -(C^{-1} \psi_1) \Gamma (\overline{\psi}_2 C) = \overline{\psi}_2 (C \Gamma^T C^{-1}) \psi_1 . \tag{II-12}$$

One may use this to show that the effect of charge conjugation on the Hamiltonian may be described as

$$\{C_i, C_i'\} \xrightarrow{C} \eta_c \{C_i^*, -C_i'^*\} . \tag{II-13}$$

Thus if the Hamiltonian is invariant under charge conjugation and not invariant under parity, then the C_i' must be pure imaginary. However, as noted before, the parity violating terms measure combinations of the form $\mathrm{Re}\, C_i C_j'^*$ and under these circumstances would vanish. This is not quite correct. If the particles that undergo weak interactions also have strong or electromagnetic interactions in the final state, then the C_i and the C_i' acquire additional phase factors which are in effect the final state scattering phase shifts. Thus in beta decay the electron and residual nucleus interact through the Coulomb potential, so that even if the C' were pure imaginary, there would be an effective $\mathrm{Re}(CC'^* e^{i\delta})$ which would be of order $\sin\delta \approx Z/137$. The fact that α is so large shows that the C' are not pure imaginary; we now know that they are real and of the same magnitude as the C_i.

* We suppress possible phase factors in this discussion which follows that of *Lee, Oehme* and *Yang* (1957).

It is interesting to ask what happens under time reversal. Here we must make the transformation

$$\psi_1 \to \overline{\psi}_1 B,$$

$$\overline{\psi}_1 \to B^{-1}\psi_1$$

where

$$BB^+ = 1\,; \qquad B(\gamma^\mu)^* B^{-1} = \gamma^\mu \tag{II-14}$$

and because of the antiunitary character of the time inversion operator (*Wigner*, 1932) we must also reverse the order of factors in any product of operators (without regard to statistics). Thus

$$\overline{\psi}_1 \Gamma \psi_2 \to (\psi_1 (B^{-1})^T \Gamma B^T \overline{\psi}_2)_{\text{reversed}} = \overline{\psi}_2 B \Gamma^T B^{-1} \psi_1 \tag{II-15}$$

and we may show that

$$\{C_i, C_i'\} \xrightarrow{T} \eta_T \{C_i^*, C_i'^*\}\,. \tag{II-16}$$

This is the basis for our earlier statement that time reversal invariance implies real coupling constants.

It is important to notice that with $\eta_P \eta_C \eta_T = 1$, a choice that we can always make, the *Hamiltonian is invariant under* the operation CPT. This result actually follows very generally for relativistic local field theories, and it is this theorem (*Lüders*, 1953; *Schwinger*, 1953; *Pauli*, 1955) and not charge conjugation invariance, for example, that requires the equality of the masses and lifetimes of particles and antiparticles.

(3) As far as testing time reversal invariance is concerned, it might appear at first sight that it is enough to measure an observable that changes sign under motion reversal, for example a quantity like $\langle \boldsymbol{S} \cdot \boldsymbol{p}_1 \times \boldsymbol{p}_2 \rangle$. That this cannot be true in general is evident from the observation of recoil proton polarisation perpendicular to the reaction plane in pion-nucleon scattering, a strong interaction for which there is good evidence that time reversal is a good symmetry. The explanation is that time reversal takes incoming states into outgoing states and if there is a final state interaction, the simple considerations no longer apply. In the beta decay of light nuclei the final state interaction is the Coulomb interaction which is not very important. One might hope that if one detects a term proportional to $\langle \boldsymbol{S} \cdot \boldsymbol{p}_e \times \boldsymbol{p}_\nu \rangle$ in the decay of polarized neutrons (*Burgy et al.*, 1958; *Erozolimsky et al.*, 1968) or Ne^{19} (*Calaprice et al.*, 1967) it would indicate a T-violation. The magnitude of the coefficient of $\langle \boldsymbol{S} \cdot \boldsymbol{p}_e \times \boldsymbol{p}_\nu \rangle$ cannot involve an interference between the C_i and the C_j' since the observable is scalar; it must vanish if all the couplings are real, and it must therefore be proportional to $\operatorname{Im} C_i C_j^*$ (and $\operatorname{Im} C_i' C_j'^*$), or more accurately, to $\operatorname{Im}(C_i C_j^* e^{i\delta})$ where δ is the Coulomb phase shift.

The measurements show no effect and yield an upper limit of about 2° for the relative phase. Thus there is no evidence for the violation of time reversal invariance in beta decay★.

(4) The statement that experiments have established the Hamiltonian (II-1) as a "serviceable approximation" is, strictly speaking incorrect. What the experiments have suggested, within the experimental errors is that the electron-neutrino current has the form $\overline{\psi}_e \gamma_\alpha (1-\gamma_5)\psi_\nu$ in beta decay. The simple extrapolation of this form to the muon-neutrino current then yields a model for the decay $\mu^\pm \to e^\pm + \nu + \nu$:

$$\mathscr{H}_1 = \frac{G}{\sqrt{2}} (\overline{\psi}_e \gamma_\alpha (1-\gamma_5)\psi_\nu)(\overline{\psi}_\nu \gamma^\alpha (1-\gamma_5)\psi_\mu) \tag{II-17}$$

and the predictions from this matrix element are in excellent agreement with experiment. One may, in fact, use this reaction to compute the coupling G, from the lifetime

$$\tau_\mu = \frac{192\,\pi^3}{G^2 m_\mu^5}. \tag{II-18}$$

Actually there are radiative corrections (Fig. 3) that must be taken into account; the laboratory measurements are carried out in a world in which the electromagnetic interactions are not turned off! The radiative corrections are finite and contribute to the electron spectrum bringing experiment and theory into closer agreement (*Sirlin*, 1969).

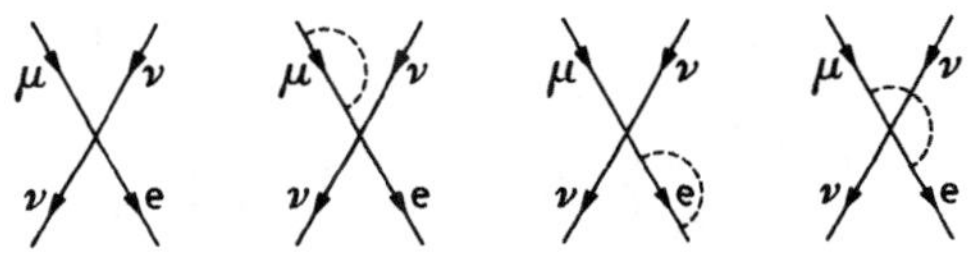

Fig. 3. Graphs that contribute to muon decay matrix element to order e^2

The finiteness of the radiative corrections is a very special feature of the form of the interaction involving $\gamma_\alpha(1-\gamma_5)$. This form has the property that it goes into itself under the Fierz rearrangement. The new matrix element now involves the form $\overline{\psi}_e \gamma_\alpha (1-\gamma_5)\psi_\mu$ which looks very much like an electric current except for the γ_5 and for the fact that the mass changes at the vertex. For the electromagnetic vertex the radiative correction is cancelled by the external line correction; each is logarithmically divergent with coefficients that do not depend on the masses. Hence the cancellation for the term $\overline{\psi}_e \gamma_\alpha \psi_\mu$ persists even though the masses

★ It should be noted that in both experiments the transition is one that connects two members of the same isomultiplet. As is pointed out in Prof. *Primakoff*'s lectures, this may be a particularly unfavorable place to look for T-violation effects.

change. The term $\overline{\psi}_e \gamma_\alpha \gamma_5 \psi_\mu$ may be written as $\overline{\psi}_{e'} \gamma_\alpha \psi_\mu$ where $\psi_{e'}$ describes an electron of mass $-m_e$. Again the terms involving the logarithm of the cut-off cancel. The result of the calculation is that if one writes (II-18) with G_{obs} in the denominator, then

$$G = G_{\text{obs}} \left[1 + \frac{\alpha}{4\pi} \left(\pi^2 - \frac{25}{4} \right) \right]. \tag{II-19}$$

Comparison with experiment yields

$$m_p^2 G = (1.020 \pm 0.001) \times 10^{-5} . \tag{II-20}$$

The Hamiltonian that we wrote down, if we restrict ourselves to leptons alone, makes some predictions about the reactions $e \to e + \nu + \nu$ and $\mu \to \mu + \nu + \nu$. One will never be able to test the latter, but there is some very indirect evidence that the former is at least compatible with astrophysical data on energy losses in stars*.

Another prediction is that the interaction Hamiltonian, together with $(H^\nu_{\text{free}} + H^e_{\text{free}} + H^\mu_{\text{free}} + H^e_{\text{e.m.}} + H^\mu_{\text{e.m.}})$ is invariant under the transformations

$$\begin{aligned} \psi_e &\to e^{i\alpha} \psi_e , \\ \psi_\mu &\to e^{i\alpha} \psi_\mu , \\ \psi_\nu &\to e^{i\alpha} \psi_\nu \end{aligned} \tag{II-21}$$

or

$$\begin{aligned} &\left.\begin{aligned} \psi_e &\to e^{i\alpha} \psi_e \\ \psi_{\nu_e} &\to e^{i\alpha} \psi_{\nu_e} \end{aligned}\right\} \quad \text{in} \quad (\overline{\psi}_e \gamma_\alpha (1 - \gamma_5) \psi_{\nu_e}) , \\ &\left.\begin{aligned} \psi_\mu &\to e^{i\beta} \psi_\mu \\ \psi_{\nu_\mu} &\to e^{i\beta} \psi_{\nu_\mu} \end{aligned}\right\} \quad \text{in} \quad (\overline{\psi}_\mu \gamma_\alpha (1 - \gamma_5) \psi_{\nu_\mu}) . \end{aligned} \tag{II-22}$$

These gauge invariance properties imply a conservation of leptons, which must be checked experimentally**. The first invariance applies when there is only one neutrino ($\nu_e \equiv \nu_\mu$); the second applies when there are two distinct neutrinos, one associated with the electron and the other with the muon. In fact, there are two neutrinos. In high energy neutrino experiments the neutrino beams are obtained from the decay of pions which overwhelmingly decay according to $\pi \to \mu + \nu_\mu$. The neutrinos interact with matter

$$\begin{aligned} \nu_\mu + p &\to l^+ + n , \\ \nu_\mu + n &\to l^- + p . \end{aligned}$$

---

* For a review and references see *Ruderman* (1969).

** Such a conservation law was first suggested by *Konopinski* and *Mahmoud* (1953).

It is observed that only μ's appear in the final state. The electrons that appear in some experiments can be explained in terms of the ν_e that come from $K^{\pm} \to \pi^0 + e^{\pm} + \nu$. This result (*Danby et al.*, 1962; *Block et al.*, 1964; *Bienlein et al.*, 1964; *Bernardini et al.*, 1964) showed that there are two neutrinos*. Furthermore if only positive pions and K's are used as a source of neutrinos, then if leptons are conserved (and we call the μ^- a lepton and a μ^+ an antilepton) then ν_μ and not $\bar{\nu}_\mu$ are produced. In the subsequent interaction with matter

$$\nu_\mu + \text{matter} \to \mu^{\pm} + \text{matter}$$

lepton conservation again predicts that only μ^- be seen. This is indeed the case (*Borer et al.*, 1968). Evidence on lepton conservation for the electron and its neutrino comes from the absence of double beta decay in which no neutrinos appear, e.g.**

$$Z^A \to (Z+2)^A + e^- + e^- .$$

Incidentally, in the two-component neutrino theory that we are using, the "neutrino" defined as the particle emitted in positron decay is lefthanded, and the "antineutrino" which is emitted in beta decay with the electron (a "lepton" and not an antilepton, by definition) is righthanded.

Let us now turn to the "semi-leptonic" beta decay matrix element. A more accurate representation than is obtainable from (II-1) is

$$\frac{G_V}{\sqrt{2}} (\bar{\psi}_e \gamma_\alpha (1-\gamma_5) \psi_{\nu_e}) \left(\bar{\psi}_p \gamma_\alpha \left(1 - \frac{G_A}{G_V} \gamma_5 \right) \psi_n \right),$$

$$G_V/G \cong 0.978 , \qquad \text{(II-23)}$$

$$G_A/G_V \simeq 1.23 \pm 0.01 .$$

The number G_V/G is determined from the lifetime of O^{14} and G_A/G_V from beta decay of the neutron; radiative corrections have been estimated. This sentence slides rather glibly over the difficulty that the radiative corrections, when calculated in the most straightforward way, are actually infinite. The reason for this is that in order to make a Fierz rearrangement that leads to a continuous "current" line, we must replace the electron operator by a positron operator. If we note that

$$\bar{\psi}_a \gamma_\alpha (1-\gamma_5) \psi_b = \bar{\psi}_b^c \gamma_\alpha (1+\gamma_5) \psi_a^c$$

we see that we now must Fierz-transform $V+A$, and this yields S and P contributions for which there is no cancellation between the vertex and

* The suggestion that there might be two neutrinos was first made by *Schwinger* (1957) and *Nishijima* (1957).

** See the lectures of *Primakoff*.

external line corrections. Thus the radiative corrections are logarithmically divergent. With any reasonable cut-off these are small and they are estimated in a reasonable way.

What Eq. (II-23) really tells us is that for low momentum transfers, the matrix element

$$\frac{G}{\sqrt{2}}\langle p|\overline{\psi}_p\gamma_\alpha(1-\gamma_5)\,\psi_n|n\rangle\,(\overline{\psi}_e\gamma^\alpha(1-\gamma_5)\,\psi_{\nu_e}) \tag{II-24}$$

with all its strong and electromagnetic renormalization effects has the form shown in (II-23). Since it is only the matrix elements that are being measured, there is really no way of checking (at low energies at least) that the *fundamental* weak interaction Hamiltonian contains neutron and proton fields. We might as well write (II-23) with

$$\begin{aligned} J_\alpha &= l_\alpha + h_\alpha\,, \\ l_\alpha &= \overline{\psi}_e\gamma_\alpha(1-\gamma_5)\,\psi_{\nu_e} + \overline{\psi}_\mu\gamma_\alpha(1-\gamma_5)\,\psi_{\nu\mu} \end{aligned} \tag{II-25}$$

and the hadronic current as yet unspecified. For such a current we have

$$\begin{aligned} \langle n|h_\alpha|p\rangle = \frac{1}{(2\pi)^3}\,&\overline{u}(n)\,\{F_1(q^2)\,\gamma_\alpha + \sigma_{\alpha\beta}q^\beta F_2(q^2) \\ &+ G_1(q^2)\,\gamma_\alpha\gamma_5 + G_2(q^2)\,q_\alpha\gamma_5\}\,u(p)\,, \\ q_\mu = n_\mu - p_\mu & \end{aligned} \tag{II-26}$$

and experiments tell us that

$$F_1(0) = 0.978\,, \qquad -G_1(0) = 1.23\,. \tag{II-27}$$

These numbers would be 1 for bare nucleons without any strong interactions. What is remarkable is that the strong interactions change these numbers (especially $F_1(0)$) so little — we do know from the size of the anomalous magnetic moments that strong interactions corrections can be large. The explanation of this remarkable phenomenon and further implications that follow from it will be discussed in the next section.

III. Currents and their Symmetries

The explanation for the very small deviation of $F_1(0)$ from unity was suggested by *Feynman* and *Gell-Mann* (1958) (see also *Gershtein* and *Zeldovich*, 1956) who noted that if the hadron currents were decomposed into vector and axial terms with changes of strangeness 0 and 1

$$h_\alpha = J_\alpha^{(\Delta S=0)} + J_{5\alpha}^{(\Delta S=0)} + J_\alpha^{(\Delta S=1)} + J_{5\alpha}^{(\Delta S=1)} \tag{III-1}$$

(with $J_\alpha^{(\Delta S=1)}$, $J_{5\alpha}^{(\Delta S=1)}$ describing leptonic decays of the strange particles), then the identification of $J_\alpha^{(\Delta S=0)}$ with the *charge-lowering part of the i-spin current* could explain $F_1(0)=1$. In the absence of electromagnetic effects the current is conserved. Furthermore

$$\int d^3x\, J_0^{(\Delta S=0)}(x) = T_- . \tag{III-2}$$

Hence

$$\begin{aligned}\int d^3x \langle n|J_0^{(\Delta S=0)}(x,0)|p\rangle &= \langle n|T_-|p\rangle \\ &= 2p_0\,\delta(\boldsymbol{p}-\boldsymbol{n}).\end{aligned} \tag{III-3}$$

On the other hand

$$\begin{aligned}\int d^3x \langle n|J_0^{(\Delta S=0)}(\boldsymbol{x},0)|p\rangle &= (2\pi)^3\,\delta(\boldsymbol{p}-\boldsymbol{n})\,\langle n|J_0^{(\Delta S=0)}(0)|p\rangle \\ &= (2\pi)^3\,\delta(\boldsymbol{p}-\boldsymbol{n})\,F_1(0)\frac{1}{(2\pi)^3}\,\overline{u}(n)\,\gamma_0 u(p) \\ &= 2p_0\,\delta(\boldsymbol{p}-\boldsymbol{n})\,F_1(0).\end{aligned} \tag{III-4}$$

From this follows

$$F_1(0)=1 . \tag{III-5}$$

This hypothesis, aside from explaining the lack of renormalization of the vector current, leads to several predictions which we shall discuss briefly.

(1) Pion beta decay. The decay

$$\pi^+ \to \pi^0 + e^+ + \nu_e$$

is governed by the matrix element

$$\langle \pi^0|J_\alpha^{(\Delta S=0)}|\pi^+\rangle = \frac{1}{(2\pi)^3}(p_{\pi^0}+p_{\pi^+})_\alpha\, f_+((p_{\pi^0}-p_{\pi^+})^2) \tag{III-6}$$

and, as in the derivation just completed we find that

$$2\omega_\pi\,\delta(\boldsymbol{p}_{\pi^0}-\boldsymbol{p}_{\pi^+})\,f_+(0) = \langle \pi^0|T_-|\pi^+\rangle$$

i.e.

$$f_+(0)=\sqrt{2}. \tag{III-7}$$

This leads to the prediction $\Gamma(\pi^+\to\pi^0 e^+\nu)/\Gamma_{tot}=1.07\times10^{-8}$. The experimental branching ratio is $(1.12\pm0.08)\times10^{-8}$ in excellent agree-

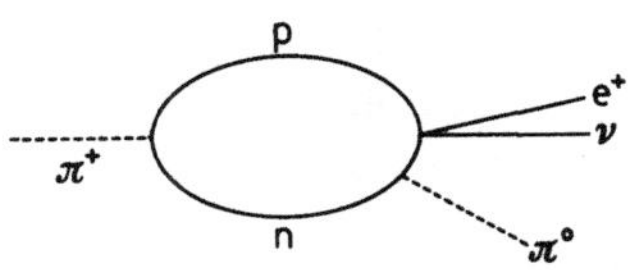

Fig. 4. Possible mechanism for pion beta decay

ment with theory. The decay would proceed with any other model (Fig. 4) but it is the merit of this hypothesis that it predicts the correct number for this branching ratio.

(2) Weak magnetism. The matrix element for the vector current between one-nucleon states may be written in the form

$$\langle p'|J_\alpha^{(\Delta S=0)}|p\rangle = \bar{u}(p')\,[\gamma_\alpha F_1(q^2) + \sigma_{\alpha\beta}q^\beta F_2(q^2)]\,\frac{\tau_1 - i\tau_2}{2}\,u(p)/(2\pi)^3\,. \quad \text{(III-8)}$$

The isovector part of the electromagnetic current, measured by subtracting the matrix elements for electron-neutron scattering from electron-proton scattering has the form

$$\begin{aligned}\langle p'|j_{\text{e.m.}\alpha}^{(V)}|p\rangle &\equiv \langle p'|J_\alpha^{(3)}|p\rangle \\ &= \bar{u}(p')\,[\gamma_\alpha F_1(q^2) + \sigma_{\alpha\beta}q^\beta F_2(q^2)]\,\frac{\tau_3}{2}\,u(p)/(2\pi)^3\end{aligned} \quad \text{(III-9)}$$

with

$$F_1(0) = 1 \quad \text{and} \quad F_2(0) = (\mu_p - \mu_n)/2M\,.$$

The identification thus tells us what the vector form factors in high energy reactions like

$$\bar{\nu}_\mu + p \to \mu^+ + n$$

are. The magnetic moment terms, proportional to $F_2(0)$ contribute to beta decay in spite of the smallness of the factor $\sigma_{\alpha\beta}q^\beta$ in the low momentum transfer beta transitions.

Gell-Mann (1958) pointed out that the weak magnetism term leads to an energy dependence in the difference between the decay spectra of B^{12} and N^{12} by e^- and e^+ emission respectively. The details of the calculation are a little too long to be presented here. The results of the experiment (for a review of theory and experiment see *Wu*, 1964) are in excellent agreement with the predictions of the CVC theory*.

Of great theoretical importance is the fact that the identification specifies in a completely *model independent way* the nature and the *scale* of the operator $J_\alpha^{(\Delta S=0)}$, the latter by the nonlinear commutation relation

$$[\int d^3x\, J_0^{(\Delta S=0)}(x,0)^+, \int d^3x'\, J_0^{(\Delta S=0)}(x',0)] = 2T_3 \quad \text{(III-10)}$$

for *i*-spin operators. In view of the identification as the charge lowering component we shall now switch notations from $J_\alpha^{(\Delta S=0)}$ to the more accurate $J_\alpha^{(1)} - iJ_\alpha^{(2)}$ or, for brevity $J_\alpha^{(1-i2)}$.

* The identification of the vector current with a component of the *i*-spin current has traditionally been abbreviated to CVC (conserved vector current).

It was, of course very tempting to generalize this model independence of the description to the rest of the hadronic currents, namely the axial currents and the strangeness changing currents. *Gell-Mann* (1962, 1964) suggested that the axial vector $J_{5\alpha}^{(\Delta S=0)}$ (which we also write as $J_{5\alpha}^{(1-i2)}$ to denote its charge-lowering property) may also be a well defined object in strong interaction physics. He suggested that in the strong interactions there exists an underlying symmetry, called *chiral symmetry*. The transformations of this symmetry, if we restrict ourselves temporarily to non-strange particles, is generated by six operators, the usual *i*-spin operators, which we will denote by Q_i, for which

$$[Q_i, Q_j] = ie_{ijk}Q_k \tag{III-11}$$

and another set of three denoted by Q_{5i}. These transform as isovectors, i.e.

$$[Q_{5i}, Q_j] = ie_{ijk}Q_{5k} \tag{III-12}$$

and furthermore

$$[Q_{5i}, Q_{5j}] = ie_{ijk}Q_k\,. \tag{III-13}$$

The group of transformations generated by the operator in the Hilbert space

$$U = \exp\{-i\boldsymbol{\alpha}\cdot\boldsymbol{Q} - i\boldsymbol{\beta}\cdot\boldsymbol{Q}_5\} \tag{III-14}$$

is conventionally called $SU(2)\times SU(2)$*. Chiral symmetry as a broken symmetry implies that the strong interactions may be described by the Hamiltonian

$$H = H_0 + \varepsilon H_1 \tag{III-15}$$

with

$$[H_0, Q_i] = [H_0, Q_{5i}] = 0\,. \tag{III-16}$$

Such a separation only makes sense if either ε is small so that we may use perturbation theory about the symmetric solutions, or if H_1 transforms like a definite tensor under the group. Most calculations carried out in the last few years assume the simplest nontrivial transformation property for H_1**.

The Q_{5i} will, in a local field theory have associated with them axial currents $J_{5\alpha}^i$, with

$$Q_{5i} = \int d^3x\, J_{50}^i(\boldsymbol{x})\,. \tag{III-17}$$

These axial currents thus are very similar to the vector *i*-spin currents, and Gell-Mann conjectured that it is these axial currents that appear in the weak interaction Hamiltonian.

---

* This is because the operators $Q_\pm = \frac{1}{2}(Q \pm Q_5)$ separately obey $SU(2)$ commutation relations and commute with each other.

** For a review of some of the work done within the framework of effective Lagrangians, see *Gasiorowicz* and *Geffen* (1969).

The axial currents are not conserved when the symmetry is broken. If they were conserved, the pion could not decay. To see this, note that quite generally

$$\langle \mu, \nu | H_1 | \pi(q) \rangle = \frac{G}{\sqrt{2}} \int d^3x \langle \mu, \nu | l_\alpha^+(x) | 0 \rangle \times \langle 0 | J_5^{(1-i2)\alpha} | \pi^+(q) \rangle \quad \text{(III-18)}$$

and

$$\langle 0 | J_5^{(1-i2)\alpha} | \pi^+(q) \rangle \equiv (2\pi)^{-3/2} i q^\alpha \sqrt{2}\, f_\pi . \quad \text{(III-19)}$$

The pion decay constant f_π is determined from the lifetime of the pion and has the experimental value of 94 MeV.

If $\partial^\alpha J_{5\alpha}^{(i)} = 0$ this would imply that

$$q^2 f_\pi = m_\pi^2 f_\pi = 0 . \quad \text{(III-20)}$$

It is in fact frequently argued that the symmetry is implemented through the existence of massless pions (Goldstone bosons) and that the pions acquire their mass because the symmetry is broken. This point of view, due to *Nambu* (1962) is very clearly propounded in recent papers by *Dashen* (1969) and *Dashen* and *Weinstein* (1969). In the real world the operator

$$D^{(i)}(x) = \partial^\alpha J_{5\alpha}^{(i)}(x) \quad \text{(III-21)}$$

does not vanish. It has the same transformation properties as the pion field and we may use it as the pion interpolating field. We thus make the identification

$$D^{(i)}(x) = m_\pi^2 f_\pi \phi_\pi^{(i)}(x) . \quad \text{(III-22)}$$

Then

$$\begin{aligned} \langle 0 | D^{(-)}(0) | \pi^{(+)} \rangle &= \langle 0 | \sqrt{2} m_\pi^2 f_\pi \phi_\pi^{(-)} | \pi^{(+)} \rangle \\ &= (2\pi)^{-3/2} \sqrt{2}\, m_\pi^2 f_\pi \end{aligned} \quad \text{(III-23)}$$

in agreement with what is expected from (III-19). In beta decay we have

$$\begin{aligned} \langle n | D^{(-)} | p \rangle &= -i(p-n)^\alpha (2\pi)^{-3} \bar{u}(n) \{\gamma_\alpha \gamma_5 G_1(q^2) + \gamma_5 q_\alpha G_2(q^2)\} u(p) \\ &= (2\pi)^{-3} \bar{u}(p)\, i\gamma_5 u(n) \, [2m G_1(q^2) + q^2 G_2(q^2)] . \end{aligned} \quad \text{(III-24)}$$

On the other hand

$$\begin{aligned} \langle n | D^{(-)} | p \rangle &= m_\pi^2 f_\pi \sqrt{2} \langle n | \phi_\pi^{(-)} | p \rangle \\ &= (2\pi)^{-3} m_\pi^2 f_\pi \sqrt{2} \bar{u}(n)\, i\gamma_5 u(p) \\ &\quad \times \left\{ \frac{g_{NN\pi}\sqrt{2}}{m_\pi^2 - q^2} + \text{non-pole terms} \right\} . \end{aligned} \quad \text{(III-25)}$$

A comparison of the two expressions shows that provided the pole term dominates some distance away from the pole (specifically at $q^2 = 0$) then

$$m G_1(0) \cong f_\pi g_{NN\pi} \,. \qquad \text{(III-26)}$$

This relation, originally discovered in a different way by *Goldberger* and *Treiman* (1958) is in good agreement with experiment. Given the value of $G_1(0)$, we find that $f^{\text{theo}} = 84$ MeV. The assumption that the pion pole dominates the matrix element of the axial current divergence has received the acronym PCAC ("partially conserved axial current") and in addition to the above application (*Gell-Mann* and *Levy*, 1960) has been very successful in other applications*.

Incidentally the pole at $q^2 = m_\pi^2$ shown in (III-25) resides in $G_2(q^2)$. This term is called the "induced pseudoscalar contribution" to the axial current matrix element. Again assuming slow variation, we have

$$G_2(q^2) \approx \frac{2 f_\pi g_{NN\pi}}{m_\pi^2 - q^2} \,. \qquad \text{(III-27)}$$

Since this term contributes to muon capture in nuclei, the assumption can be tested**.

The identification of the axial current with the generating current of $SU(2) \times SU(2)$ leads to a relation, via the PCAC condition, between the axial renormalization constant (G_A/G_V) and some strong interaction parameters. This relation due to *Adler* (1965) and *Weisberger* (1965) is in excellent agreement with experiment. The derivation of that relation is a little too long to be presented here. In the briefest of outlines, the starting point is an identity of the form $(p + q = p' + q')$

$$\begin{aligned} q'_\mu q_\nu \int \mathrm{d}x\, \mathrm{e}^{iq'x} \langle p' | T(J_5^{\mu(\alpha)}(x)\, J_5^{\nu(\beta)}(0)) | p \rangle & \\ = \int \mathrm{d}x\, \mathrm{e}^{iq'x} \langle p' | T(D^{(\alpha)}(x)\, D^{(\beta)}(0)) | p \rangle & \\ + q_\nu\, \mathrm{e}^{\alpha\beta\gamma} \langle p' | J^{\nu(\gamma)}(0) | p \rangle & \\ - \langle p' | \sigma^{\alpha\beta}(0) | p \rangle & \end{aligned} \qquad \text{(III-28)}$$

obtained by differentiating the step functions in the T-product. The term $\sigma^{\alpha\beta}$ is defined by

$$[D^{(\beta)}(x), J_{50}^{(\alpha)}(0)]_{x_0=0} = \delta(x)\, \sigma^{\alpha\beta}(0) \qquad \text{(III-29)}$$

and the second term on the r.h.s. comes from

$$[J_{50}^{(\alpha)}(x), J_{5\nu}^{(\beta)}(0)]_{x_0=0} = i e_{\alpha\beta\gamma} J_\nu^{(\gamma)}(0)\, \delta(x) \,. \qquad \text{(III-30)}$$

* See the monographs of *Adler* and *Dashen* (1968) and *Renner* (1968) for some of the many applications.

** See the lectures of *Primakoff*.

Actually, when $\nu \neq 0$ there are further terms on the right hand side: these are the so-called "Schwinger terms" (*Goto* and *Imamura*, 1955; *Schwinger*, 1959). Furthermore the T-product of two currents is generally not a tensor — to make it so one needs to add certain other terms to it. It is accepted practice to ignore both difficulties simultaneously since in certain models they are known to cancel each other out. The $\sigma^{\alpha\beta}$ term is symmetric in (α, β) in most models of interest. Thus in considering the antisymmetric part of (III-28) we can ignore it. Now in the limit $q, q' \to 0$ the only contributions to the left hand side of our identity come from the terms in the T-product that are singular in the limit. These, as one can easily convince oneself by drawing some Feynman graphs, only involve the one-nucleon intermediate states. Thus the coupling of axial currents to nucleons enters, and G_A/G_V makes its appearance. On the right hand side, the current matrix element is known. The first term on the r.h.s. may be related to a particular off-shell extrapolation of the pion-nucleon scattering matrix element. The latter is defined as follows

$$T_{\beta\alpha}(p', q'; p, q) = -\frac{i}{(2\pi)^3} \lim_{\substack{q^2 = m_\pi^2 \\ q'^2 = m_\pi^2}} \int dx\, e^{iq'x} \qquad \text{(III-31)}$$
$$(m_\pi^2 - q^2)(m_\pi^2 - q'^2) \langle p' | T(\phi_\beta(x)\, \phi_\alpha(0)) | p \rangle$$

and in the off-shell extrapolation we just leave off the limits and use (III-22). Straightforward manipulations lead to the Adler-Weisberger relation (*Weinberg*, 1966). The rather subtle problems connected with estimating errors in going to the physical region in which pions are not massless have been discussed by *Fubini* and *Furlan* (1968).

With this very suggestive identification of the strangeness preserving vector and axial currents, the step to enlarge the algebra to encompass the strangeness changing currents was made easier by the recent discovery* that $SU(3)$ was an approximate symmetry of the strong interactions. Thus the enlargement of chiral symmetry to $SU(3) \times SU(3)$ led to the definition of *eight* vector and axial currents. These currents have the same transformation properties (as far as the internal variables are concerned) as the $SU(3)$ generators. To maintain charge conservation the strangeness changing currents must also be charge-lowering currents. The only charge lowering currents transform as $Q^{(1)} - iQ^{(2)}$ (these do not change strangeness) and $Q^{(4)} - iQ^{(5)}$. We thus take for the $\Delta S = 1$ vector and axial currents $J_\alpha^{(4)} - iJ_\alpha^{(5)}$ and $J_{5\alpha}^{(4)} - iJ_{5\alpha}^{(5)}$ respectively.

The strangeness changing currents have the property that $\Delta T = 1/2$ and also $\Delta S = \Delta Q$. The evidence for the first selection rule is still somewhat

* See *Gell-Mann* and *Ne'eman* (1964).

scant. The main prediction is that the ratios

$$\frac{\Sigma\Gamma(K_L\to\pi^\pm e^\mp \nu)}{2\Gamma(K^+\to\pi^0 e^+ \nu)}\,,$$

$$\frac{\Sigma\Gamma(K_L\to\pi^\pm \mu^\mp \nu)}{2\Gamma(K^+\to\pi^0 e^+ \nu)}$$

should both equal unity within 2% coming from radiative corrections and mass differences. The ratio for the electrons is 0.99 ± 0.04 and that for muons is 1.05 ± 0.04 (*Rubbia*, 1969) in excellent agreement with theory. There is at this time no evidence for the violation of the $\Delta S=\Delta Q$ rule. For example, the reaction

$$\Sigma^-\to n+e^-+\bar{\nu}_e$$

has frequently been observed, while no known cases of

$$\Sigma^+\to n+e^++\nu_e$$

exist. More quantitatively

$$\frac{\Gamma(\Sigma^+\to ne^+\nu)}{\Gamma(\Sigma^-\to ne^-\nu)}<1.7\%\,.$$

Similarly no cases of the forbidden reaction

$$K^+\to\pi^+\pi^+e^-\bar{\nu}_e$$

have been seen, whereas the reaction

$$K^+\to\pi^+\pi^-e^+\nu_e$$

has frequently been seen to occur. The study of the decay

$$K^0\to\pi^+e^-\bar{\nu}_e$$

which is supposed to be forbidden, shows that if we write

$$\frac{A(K^0\to\pi^+e^-\bar{\nu}_e)}{A(K^0\to\pi^-e^+\nu_e)}=x$$

then present experiments yield

$$\begin{aligned}\mathrm{Re}\,x &= 0.14\pm0.05\,,\\ \mathrm{Im}\,x &= -0.12\pm0.15\,,\\ \frac{1-|x|^2}{|1-x|^2} &= 1.06\pm0.06\end{aligned}$$

(*Filthuth*, 1969). If there were $\Delta S = -\Delta Q$ currents, then the nonleptonic interactions should allow $\Delta S = 2$ transitions. No cases of

$$\Xi^- \rightarrow n + \pi^-$$

have been seen. Furthermore the "mixing" rate of K^0 and $\bar{K}^0$ is known to be of second order in the weak coupling constant G, i.e. there is no appreciable $\Delta S = 2$ interaction.

If our conjectures about the nature of the strangeness changing current are correct, all that remains is to determine the relative amount of $\Delta S = 1$ and $\Delta S = 0$ currents in J_α. If we compare $\Gamma(K \rightarrow \mu\nu)$ and $\Gamma(\pi \rightarrow \mu\nu)$, we see that the ratio of the rates may be expressed in terms of a known phase space factor and the ratio $(f_K/f_\pi)^2$ where

$$\langle 0 | J^{(4)}_{5\mu} - iJ^{(5)}_{5\mu} | K^+ \rangle = (2\pi)^{-3/2} i q^{(K)}_\mu \sqrt{2} f_K .$$

If $SU(3)$ were a good symmetry this ratio would be one. To fit the experiments we need

$$\left| \frac{f_K}{f_\pi} \right| \simeq 0.25 .$$

Since we do not expect $SU(3)$ symmetry breaking to be so large, we expect that this ratio rather reflects a difference in the coefficients in front of the currents. A better test is a comparison of the rates $\Gamma(K^+ \rightarrow \pi^0 e^+ \nu_e)$ and $\Gamma(\pi^+ \rightarrow \pi^0 e^+ \nu_e)$. The matrix element for the K decay has the form

$$\langle \pi^0 | J^{(4)}_\mu - iJ^{(5)}_\mu | K^+ \rangle = \frac{1}{(2\pi)^3} [f_+((q_K - q_\pi)^2)(q_K + q_\pi)_\mu + f_-((q_K - q_\pi)^2)(q_K - q_\pi)_\mu] . \quad \text{(III-32)}$$

Since this matrix element is to be multiplied by the leptonic current, one finds that the second term in the expression is suppressed by a factor $(m_e/m_K)^2$ and we ignore it in our discussion. In the $SU(3)$ limit the vector current is a pure F current and thus the expected ratio is

$$\left(\frac{f^K_+(0)}{f^\pi_+(0)} \right)^2 = \frac{1}{4} . \quad \text{(III-33)}$$

The expection that this is correct is much larger than in the case of the axial decay because of the theorem of *Ademollo* and *Gatto* (1964) which states that the departure from this formula are of second order in $SU(3)$ breaking. The experimental ratio is of the order 0.01 indicating a suppression of the strangeness changing current. Similar experimental suppression appears in the rate for Λ beta decay and other semileptonic decays that show a change of strangeness.

A formulation of such a suppression which maintains in a very satisfactory way the notion of universality is due to *Cabibbo* (1964), (see also *Gell-Mann*, 1964). The notion of universality may be formed by requiring that the *charges of the weak currents generate an* $SU(2)$ *algebra.* The weak electronic charge is $\int d^3x\, \bar{\psi}_e(x)\, \gamma_0(1-\gamma_5)\, \psi_{\nu_e}(x)$ etc. For the leptons we use canonical commutation relations and find that this postulate is satisfied, provided the electron and muon currents enter in equal amounts. For the hadronic currents we find that the current must be of the form

$$\begin{aligned} h_2 &= e^{-2iF_7\theta}(J_\mu^{(1-i2)} + J_{5\mu}^{(1-i2)})\, e^{2iF_7\theta} \\ &= \cos\theta(J_\mu^{(1-i2)} + J_{5\mu}^{(1-i2)}) \\ &\quad + \sin\theta(J_\mu^{(4-i5)} + J_{5\mu}^{(4-i5)})\,. \end{aligned} \tag{III-34}$$

The new parameter, the *Cabibbo angle* θ is the same for the vector and the axial current. Indeed an examination of the leptonic decays of hyperons shows that

$$\theta_V = 0.233 \pm 0.012\,,$$
$$\theta_A = 0.238 \pm 0.018$$

(see *Filthuth*, 1969). The decay $K^+ \to \pi^0 e^+ \nu_e$ yields*

$$\sin\theta = 0.220 \pm 0.003\,. \tag{III-35}$$

One then predicts that

$$\frac{f_K}{f_\pi} = 1.23 \pm 0.05 \tag{III-36}$$

which is consistent with experience in other areas of the strong interactions where the deviations from $SU(3)$ for the coupling constants tend to be small.

As can be seen from (III-34), the Cabibbo angle is a rotation angle between two directions in $SU(3)$ space. One of them is the direction of the *i*-spin axis which is "picked out" by the medium strong $SU(3)$ symmetry violation; the other is the direction of the violation due to the weak interactions. If θ were zero, for example, strange particles would not decay; however we only know "strange" particles from "nonstrange" ones because $SU(3)$ is violated even by the strong interactions. To understand the magnitude of the Cabibbo angle, we must understand the relation between these two kinds of symmetry breakings. Recent ideas on this subject will be discussed by Prof. *Gatto*.

* We assume here that $f_+^K(0) = 1$. Calculations with chiral Lagrangians support the Ademollo-Gatto theorem and show that $f_+^K(0)$ is reduced from unity by 2–4%.

Recent progress in the understanding of chiral symmetry through current algebras has led to an increased understanding of the form factors that appear in the weak semileptonic decays. Let us illustrate this with a derivation of a relation between the axial and the vector form factors in K decay (*Callan* and *Treiman* (1966); *Mathur*, *Okubo*, and *Pandit* (1966)). Let us consider the matrix element in (III-32), with a change in notation: the form factors will be written as a function of its three variables $f_\pm(q_K^2, q_\pi^2, (q_K-q_\pi)^2)$. The pion is now contracted by the reduction procedure of *Lehmann*, *Symanzik*, and *Zimmermann* (1955) and we obtain

$$\langle \pi^0 | J_\mu^{(4-i5)} | K^+ \rangle = i(m_\pi^2 - q_\pi^2) \int dx\, e^{iq_\pi x} \langle 0 | T(\phi^{(3)}(x), J_\mu^{(4-i5)}(0)) | K^+ \rangle / (2\pi)^{3/2}$$

which may be rewritten as

$$i\frac{m_\pi^2 - q_\pi^2}{m_\pi^2 f_\pi} \int dx\, e^{iq_\pi x} \langle 0 | T(\partial^\lambda J_{5\lambda}^{(3)}(x), J_\mu^{(4-i5)}(0)) | K^+ \rangle / (2\pi)^{3/2}$$

$$= \frac{m_\pi^2 - q_\pi^2}{m_\pi^2 f_\pi} q_\pi^\lambda \int dx\, e^{iq_\pi x} \langle 0 | T(J_{5\lambda}^{(3)}(x), J_\mu^{(4-i5)}(0)) | K^+ \rangle / (2\pi)^{3/2}$$

$$- i\frac{m_\pi^2 - q_\pi^2}{m_\pi^2 f_\pi} \int dx\, e^{iq_\pi x} \delta(x_0) \langle 0 | [J_{50}^{(3)}(x), J_\mu^{(4-i5)}(0)] | K^+ \rangle / (2\pi)^{3/2}.$$

Let us now take the soft pion limit $q_\pi \to 0$ and use the commutation relations of $SU(3) \times SU(3)$. The first term vanishes because there are no singular terms in the T-product. We then get

$$q_{K\mu}(f_+(m_K^2, 0, m_K^2) + f_-(m_K^2, 0, m_K^2))/(2\pi)^3$$

$$= -\frac{i}{(2\pi)^{3/2} f_\pi} \langle 0 | [Q_5^{(3)}, J_\mu^{(4-i5)}] | K^+ \rangle$$

$$= -\frac{1}{(2\pi)^3} \frac{1}{\sqrt{2}} \frac{f_K}{f_\pi} q_{K\mu}.$$

Comparing this with the $SU(3)$ limit

$$\langle \pi^0 | J_\mu^{(4-i5)} | K^+ \rangle = \frac{1}{(2\pi)^3} (q_K + q_\pi)_\mu \left(-\frac{1}{\sqrt{2}}\right)$$

we find that the CTMOP formula reads

$$f_+(m_K^2, 0, m_K^2) + f_-(m_K^2, 0, m_K^2) = \frac{f_K}{f_\pi}. \tag{III-37}$$

In the symmetry limit the first and third factor are unity and the second one vanishes.

Although this formula does not provide us with all the information about K_{l3} decays, it does provide an additional constraint on the dynamical model that is used to study the matrix element. There is much literature on the subject and some of it will surely be discussed by Dr. *Müller* and Dr. *Renner*.

IV. Nonleptonic Decays

The (current) × (current) form of the weak interaction Hamiltonian predicts that there should exist weak nonleptonic interactions to order G. They should be parity violating, PC conserving and have $\Delta S = 0$ and 1 — and that is about all that we can confidently deduce from the form of the Hamiltonian! The $\Delta S = 0$ nonleptonic interactions, e.g.

$$n + p \xrightarrow{\text{weak}} n + p$$

will always be completely overshadowed by strong and electromagnetic interactions that also conserve strangeness. The only hope of detecting these weak interactions is to observe parity violations in nuclear γ decay, caused by the parity mixing of the nuclear levels. The experimental situation will be discussed by Prof. *Telegdi*. The $\Delta S = 1$ interactions have been studied in great detail. Those that studied in some detail are

(a) $\Lambda^0 \to p + \pi^-$,
$\Lambda^0 \to n + \pi^0$.

(b) $\Sigma^+ \to p + \pi^0$,
$\to n + \pi^+$,
$\Sigma^- \to n + \pi^-$.

(c) $\Xi^- \to \Lambda^0 + \pi^-$,
$\Xi^0 \to \Lambda^0 + \pi^0$.

(d) $K^\pm \to \pi^\pm + \pi^0$,
$K^0 \to \pi^+ + \pi^-$,
$K^0 \to \pi^0 + \pi^0$.

(e) $K^\pm \to \pi^\pm + \pi^+ + \pi^-$,
$\to \pi^\pm + \pi^0 + \pi^0$,
$K^0 \to \pi^+ + \pi^- + \pi^0$.

Let us start with a phenomenological analysis of the hyperon decays. For the decays of spin 1/2 particles, the most general form of the matrix element is

$$\langle B_f(p_f), \pi | H_w | B_i(p_i)\rangle = \frac{1}{(2\pi)^3} \bar{u}(p_f)\,[A + B\gamma_5]\,u(p_i)\,. \qquad \text{(IV-1)}$$

The amplitudes will have an additional labeling, e.g. $A(\Sigma_0^+)$ denotes the S-wave amplitude for the decay of a Σ^+ into a π^0. The S-wave amplitudes are parity violating ones (parity conservation requires the pseudoscalar mesons to couple through a γ_5) and the P-wave amplitudes $B(\ldots)$ are the parity conserving ones.

The seven reactions (a)–(c) thus are described by fourteen amplitudes. If CP is a good symmetry – and the evidence is that it is – then the phases of the A and B amplitudes are the $J = 1/2$ S and P wave scattering phase shifts for the two particles in the final state, i.e. pion-nucleon phase shifts in the first five reactions and pion $-\Lambda^0$ in the last two. The phase shifts are to be evaluated at the center of mass energy corresponding to the mass of the decaying particle. In the first five reactions the phases are known and are small. We shall therefore ignore them. (IV-1) may be rewritten in the form

$$\langle B_f(p_f), \pi | H_w | B_i(p_i)\rangle = \frac{1}{(2\pi)^3}(a + b\boldsymbol{\sigma}\cdot\hat{\boldsymbol{p}}_f)\,. \qquad \text{(IV-2)}$$

With the help of the standard representation of the Dirac spinors

$$u_i(\boldsymbol{p}_i = 0) = \sqrt{2m_i}\begin{pmatrix}1\\0\end{pmatrix}; \qquad u_f(\boldsymbol{p}_f) = \sqrt{E_f + m_f}\begin{pmatrix}1\\ \dfrac{\boldsymbol{\sigma}\cdot\boldsymbol{p}_f}{E_f + m_f}\end{pmatrix}$$

we see that

$$a = [2m_i(E_f + m_f)]^{1/2} A\,,$$
$$b = -[2m_i(E_f - m_f)]^{1/2} B\,.$$

The decay rate measures the spin average of the absolute square of (IV-2); thus from the decay rate we obtain. If the initial baryon is polarized, with polarization P_i in the direction $\hat{n}_i$, and the final baryon has polarization P_f in the direction $\hat{n}_f$, then the rate is proportional to

$$\begin{aligned}
\tfrac{1}{4}\,\mathrm{tr}\,&((a + b\boldsymbol{\sigma}\cdot\hat{\boldsymbol{p}}_f)(1 + P_i\boldsymbol{\sigma}\cdot\hat{\boldsymbol{n}}_i)(a^* + b^*\boldsymbol{\sigma}\cdot\hat{\boldsymbol{p}}_f)(1 + P_f\boldsymbol{\sigma}\cdot\hat{\boldsymbol{n}}_f))\\
&= \tfrac{1}{2}[|a|^2 + |b|^2 + 2P_i\hat{\boldsymbol{p}}_f\cdot\hat{\boldsymbol{n}}_i\,\mathrm{Re}(ab^*)\\
&\quad + 2P_f\hat{\boldsymbol{p}}_f\cdot\hat{\boldsymbol{n}}_f\,\mathrm{Re}(ab^*) + P_iP_f\hat{\boldsymbol{n}}_i\cdot\hat{\boldsymbol{n}}_f(|a|^2 - |b|^2) \qquad \text{(IV-3)}\\
&\quad + 2P_iP_f\hat{\boldsymbol{p}}_f\cdot\hat{\boldsymbol{n}}_i\hat{\boldsymbol{p}}_f\cdot\hat{\boldsymbol{n}}_f|b|^2\\
&\quad + 2P_iP_f\,\mathrm{Im}(ab^*)\,\hat{\boldsymbol{p}}_f\cdot\hat{\boldsymbol{n}}_i\times\hat{\boldsymbol{n}}_f]\,.
\end{aligned}$$

Thus various polarization measurements can determine the magnitude and relative phase of a and b. The last term will be very tiny if CP invariance holds, since the term $\mathrm{Im}(ab^*)$ involves only the tiny phase shifts mentioned before. A large correlation there would prove T-invariance violation, but such a correlation has not been observed. The current data from the best studied Λ^0 decay is

$$\left[\frac{\mathrm{Im}(ab^*)}{\mathrm{Re}(ab^*)}\right]_{\mathrm{theo}} = -0.11 \pm 0.03\,; \qquad \left[\frac{\mathrm{Im}(ab^*)}{\mathrm{Re}(ab^*)}\right]_{\mathrm{exp}} = -0.13 \pm 0.06\,.$$

Various values of the parameters may be found in the report by *Filthuth* (1969).

In discussing the phases of a and b I did not specify, as I should have done, whether the phases were in the $T=1/2$ or in the $T=3/2$ state. The correct choice is $T=1/2$ because of the existence of a remarkably accurate empirical law which states that the *various nonleptonic transition matrix elements transform like isospinors* to a very good approximation, i.e. of the possible allowed changes $\Delta T = 1/2, 3/2, 5/2, \ldots$, the transitions $\Delta T = 1/2$ dominate all others (*Gell-Mann* and *Pais*, 1954).

The rule was first postulated in connection with the $K_{2\pi}$ decays. The rates for the two processes

$$K^0 \to \pi^+ + \pi^-\,,$$

$$K^+ \to \pi^+ + \pi^0$$

differ by a factor of 500. The only difference in the system is that in the neutral decay the symmetric two-pion state ($J=0$) can be in a $T=0$ as well as a $T=2$ state, whereas the state can only have $T=2$. A $\Delta T = 1/2$ enhancement could thus explain the difference. The fact that the K^+ does decay shows that the $\Delta T = 3/2$ matrix element does not vanish. It is supressed by something of the order of 20 which is too large a suppression to be accidental, but is not large enough if the $\Delta T = 1/2$ rule were only violated by electromagnetic effects (which are of order α).

The predictions of the rule for hyperon decays are easily obtained by treating H_w like a "vertex" for the emission of an i-spin 1/2 object. Thus

(a) in the case of Λ^0 decay, described by

$$\underbrace{\text{“spurion”} + \Lambda^0}_{T=1/2} \begin{cases} \nearrow p + \pi^- \equiv (\psi_{3/2} - \sqrt{2}\,\psi_{1/2})/\sqrt{3} \\ \searrow n + \pi^0 \equiv (\sqrt{2}\,\psi_{3/2} + \psi_{1/2})/\sqrt{3} \end{cases}$$

the ratio

$$\frac{A(\Lambda_-^0)}{A(\Lambda_0^0)} = \frac{B(\Lambda_-^0)}{B(\Lambda_0^0)} = -\sqrt{2} \tag{IV-4}$$

is predicted. This ratio is in excellent agreement with experiment (*Filthuth*, 1969) as is a similar relation between the two modes of the $\Xi^{-,0}$ decay.

(b) The three amplitudes for the $\Sigma^{\pm}$ decay, $M(\Sigma_+^+)$, $M(\Sigma_0^+)$ and $M(\Sigma_-^-)$ may be written in the form

$$M(\Sigma_+^+) = \langle n\pi^+ | \Sigma^+, \text{spurion}\rangle = \frac{1}{3} M_{3/2} + \frac{2}{3} M_{1/2}$$

$$M(\Sigma_0^+) = \langle p\pi^0 | \Sigma^+, \text{spurion}\rangle = \frac{\sqrt{2}}{3} M_{3/2} - \frac{\sqrt{2}}{3} M_{1/2},$$

$$M(\Sigma_-^-) = \langle n\pi^- | \Sigma^-, \text{spurion}\rangle = M_{3/2}$$

so that

$$\sqrt{2}\, M(\Sigma_0^+) + M(\Sigma_+^+) = M(\Sigma_-^-). \tag{IV-5}$$

These relations hold for the S and P waves separately, so that if we plot an $S-P$ plane (and treat the amplitudes as real) they should form a closed triangle. Again the agreement is very good.

The Hamiltonian H_w does not transform like a $T=1/2$ object, since out of the $\Delta T=1$, $\Delta S=0$ current and out of the $\Delta T=1/2$, $\Delta S=1$ current a term with $\Delta T=3/2$ can also be formed. One could add to the Hamiltonian that we have been dealing with purely nonleptonic terms involving neutral currents, arranged in such a way that the $\Delta T=3/2$ terms are cancelled. There is no real argument against this except an aesthetic one (one does know that there are no neutral lepton currents) and the fact that the selection rule is not perfect, so that it is possible to argue that it is the strong interactions that act to enhance the $\Delta T=1/2$ matrix elements, just as they appear to act towards the octet enhancement manifested by the Gell-Mann-Okubo mass formula. Once it is assumed that H_w behaves like an octet under $SU(3)$, it is possible to derive one more relation

$$2A(\Xi_-^-) - A(\Lambda_-^0) = \sqrt{3}\, A(\Sigma_0^+). \tag{IV-6}$$

It turns out that not only this relation (*Gell-Mann*, 1964; *Lee*, 1964; *Sugawara*, 1964), but the corresponding P-wave relation, for which there is no convincing derivation, is quite well satisfied.

We may now ask what current algebra and soft pion techniques tell us about the nonleptonic decays of the hyperons. Proceeding in a fashion that is by now familiar we obtain

$$\begin{aligned}
&\lim_{q\to 0} \langle B_f, \pi^a(q)|\mathscr{H}_w(0)|B_i\rangle \\
&\quad = \lim_{q\to 0} \frac{i(m_\pi^2-q^2)}{m_\pi^2 f_\pi} \int \mathrm{d}x\, \mathrm{e}^{iqx} \langle B_f| T(\partial^\lambda J_{5\lambda}^{(a)}(x), \mathscr{H}_w(0))|B_i\rangle \\
&\quad = \lim_{q\to 0} \frac{i(m_\pi^2-q^2)}{m_\pi^2 f_\pi} [-iq^\lambda \int \mathrm{d}x\, \mathrm{e}^{iqx} \langle B_f| T(J_{5\lambda}^{(a)}(x), \mathscr{H}_w(0))|B_i\rangle \\
&\qquad + \int \mathrm{d}x\, \mathrm{e}^{iqx} \delta(x_0) \langle B_f|[J_{50}^{(a)}(x), \mathscr{H}_w(0)]|B_i\rangle \\
&\quad = \frac{i}{f_\pi} \langle B_f|[Q_5^{(a)}, \mathscr{H}_w(0)]|B_i\rangle \\
&\qquad + \lim_{q\to 0} \frac{q^\lambda}{f_\pi} \int \mathrm{d}x\, \mathrm{e}^{iqx} \langle B_f| T(J_{5\lambda}^{(a)}(x), \mathscr{H}_w(0))|B_i\rangle\,.
\end{aligned} \tag{IV-7}$$

In the second term only the single baryon pole terms will contribute, since they can be singular at $q=0$. The contributions involve the unknown $\langle B_n|\mathscr{H}_w(0)|B_i\rangle$. If we consider the parity-violating part of H_w, for which this matrix element vanishes, we have the soft-pion theorem

$$\lim_{q\to 0} \langle B_f, \pi^a(q)|\mathscr{H}_w^{(-)}(0)|B_i\rangle = \frac{i}{f_\pi} \langle B_f|[Q_5^{(a)}, \mathscr{H}_w^{(-)}(0)]|B_i\rangle\,. \tag{IV-8}$$

It follows from the chiral properties of $\mathscr{H}_w$ that

$$[Q_5^{(a)}, \mathscr{H}_w^{(-)}(0)] = [Q^{(a)}, \mathscr{H}_w^{(+)}(0)]\,. \tag{IV-9}$$

From this one deduces that in the soft pion limit

$$A(\Sigma_+^+) + \sqrt{2}\, A(\Sigma_0^+) = A(\Sigma_-^-)\,. \tag{IV-10}$$

The pions are not, in fact, soft and model dependent corrections can be made. This subject as well as the treatment of the P-wave nonleptonic decays is discussed in the lectures of Prof. *Stech*. We add only one comment, which is that the experimental fact that

$$A(\Sigma_+^+) \cong 0 \tag{IV-11}$$

can be derived once the octet transformation property of H_w is assumed. The experimental fact that

$$B(\Sigma_-^-) \cong 0 \tag{IV-12}$$

cannot be so derived. It is clear that in the nonleptonic even more than in the semileptonic weak interactions, the difficulty lies with our inability to perform strong interaction calculations.

V. Concluding Remarks

In spite of the difficulties that we have repeatedly drawn attention to, we appear to have a remarkably coherent picture of the weak interactions at low energies. An enormous variety of phenomena are described by a very simple universal interaction, and whenever calculations are possible, they seem to agree with experiment. The picture cannot be quite correct as is seen from the fact that the cross section for a reaction like

$$e^- + \nu_e \to e^- + \nu_e$$

or

$$\bar{\nu}_\mu + p \to n + \mu^+$$

rises linearly with energy at high energies and this violates unitarity. For a point interaction, we have scattering in S states only, and the requirement that the scattering amplitude F obey

$$|F| < \frac{1}{q} \cong \frac{2}{W}$$

while from the theory

$$|F| \sim GW$$

implies that at energies for which $GW^2 \cong 1$ (W is the center of mass energy), a modification is required.

One may hope to escape this trouble by postulating that the low energy H_w is just a reflection of a more "fundamental" interaction, very similar in structure to the electromagnetic interaction, in which "non-hadronic" but very heavy vector mesons $W^\pm$ are involved

$$H_1 = g_w \int d^3x (J^\alpha(x)\, W_\alpha^+(x) + \text{h.c.}) \tag{V-1}$$

with

$$\frac{g_w^2}{m_w^2} = \frac{G}{\sqrt{2}}. \tag{V-2}$$

With this model the reactions just mentioned are described by graphs like those that appear in Moller scattering, and there is no immediate trouble with unitarity. That is unfortunately all that is gained by such a model. High energy predictions, in which higher order W exchanges occur are not well defined because the theories involving charged vector mesons are not renormalizable, i.e. divergences appear in each order, and these divergences cannot be absorbed in terms of a mass, a coupling constant or some other measurable parameters*.

No W mesons have been seen. A mechanism in which they could have been observed is shown in Fig. 5. The absence of such events in the neutrino experiments at existing energies, together with estimates of

* Current efforts to avoid some of the most obvious difficulties are discussed in the lectures of *Segre*.

their production cross section indicate that if W mesons exist, their mass must exceed 2 GeV. W mesons, like quarks will be looked for at higher energies, but it may well be that the unitarity troubles with the weak interactions may be solved in a much more profound way.

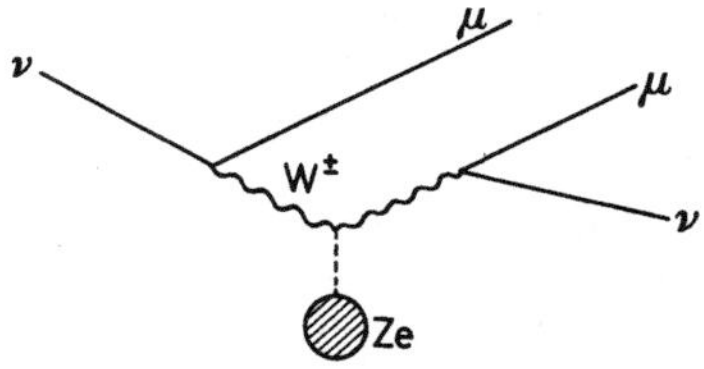

Fig. 5. Event indicating the existence of W. The muon pair production must occur with a rate of order G^2

There is another thing that is wrong with our picture of the weak interactions, and that is the fact that it does not say anything about the small degree of CP violation observed in the neutral K decays (see the lectures of *Kabir* and *Steinberger*). It may be that this phenomenon has nothing to do with the weak interactions, except that they show up a phenomenon which belongs to a different interaction. A popular belief is that there exists another "*superweak*" interaction, which is 10^8 weaker than the weak interaction which violates CP and allows $\Delta S = 2$ transitions. Such an interactions manifests itself at the present level of experimental accuracy only in the mixing that it induces in the CP even and CP odd states that form the short and long-lived neutral K mesons*. If experiments continue to bear out this isolation of the CP violation in another, inaccessible, area of physics, we may indeed be at a point of understanding the weak interactions. At the same time the hope of a unified picture of all of elementary particle physics emerging from what we now know will suffer a serious blow.

References

Ademollo, M., Gatto, R.: Phys. Rev. Letters **13**, 264 (1964).
Adler, S. L.: Phys. Rev. Letters **14**, 1051 (1965).
Allen, J. S.: Rev. Mod. Phys. **31**, 791 (1959).
Bergkvist, K. E.: Proceedings of Topical Conference on Weak Interactions, CERN (1969).
Bernardini, G., Bienlein, J. K., von Dardel, G., Faissner, H., Ferrero, F., Gaillard, J.-M., Gerber, H. J., Hahn, B., Holder, M., Kaftanov, V., Krienen, F., Manfredotti, C., Reinharz, M., Salmeron, R. A., Staude, A., Steiner, H. J., Bartley, J., Block, M. M., Burmeister, H., Cundy, D. C., Eibern, B., Franzinetti, C., Keren, J., Møllerud, H., R., Myatt, G., Nikolic, M., Orkin-Lecourtois, A., Paty, M., Perkins, D. H., Ramm, C. A., Schultze, K., Sletten, H., Soop, K., Stump, R., Venus, W., Yoshiki, H.: Nuovo Cimento **38**, 608 (1965).

* An argument in favor of this hypothesis is the fact that the measurements of the electric dipole moment of the neutron is $(2 \pm 2) \times 10^{-23}$ e cm (*Miller et al.*, 1968). Theoretical estimates would put it at 10^{-21} e cm.

Bienlein, J. K., Bohm, A., von Dardel, G., Faissner, H., Ferrero, F., Gaillard, J.-M., Gerber, H. J., Hahn, B., Kaftanov, V., Krienen, F., Reinharz, M., Salmeron, R. A., Seiler, P. G., Staude, A., Stein, J., Steiner, H. J.: Phys. Rev. Letters **13**, 80 (1964).
Block, M. M., Burmeister, H., Cundy, D. C., Eiben, B., Franzinetti, G., Keren, J., Møllerud, R., Maytt, G., Nikolic, M., Orkin-Lecourtois, A., Paty, M., Perkins, D., Ramm, C. A., Schultz, K., Sletten, H., Soop, K., Stump, R., Venus, W., Yoshiki, H.: Phys. Rev. Letters **12**, 281 (1964).
Borer, K., Hahn, B., Hofer, H., Kaspar, H., Krienen, F., Seiler, T.-G.: Proceedings of the Vienna Conference (1968).
Burgy, M. T., Krohn, V. E., Novey, T. B., Ringo, G. K., Telegdi, V. L.: Phys. Rev. **110**, 1214 (1958); — Phys. Rev. Letters **1**, 324 (1958).
Cabibbo, N.: Phys. Rev. Letters **10**, 531 (1963).
Calaprice, F. P., Commins, D. E., Gibbs, H. M., Wick, G. L., Dobson, D. A.: Phys. Rev. Letters **18**, 918 (1967).
Callan, C., Treiman, S. B.: Phys. Rev. Letters **16**, 153 (1966).
Dalitz, R.: Phil. Mag. **44**, 1068 (1953); — Phys. Rev. **94**, 1046 (1954).
Danby, G., Gaillard, J. M., Goulianos, K.,Lederman, L. M., Mistry, N., Schwartz, M., Steinberger, J.: Phys. Rev. Letters **9**, 36 (1962).
Dashen, R.: Phys. Rev. **183**, 1245 (1969).
— *Weinstein, M.:* Phys. Rev. **183**, 1261 (1969).
Erozolimsky, S. G., Bondarenko, L. N., Mostovoy, Yu. A., Obinyakov, B. A., Zacharova, V. P., Titov, V. A.: Phys. Letters B **27**, 557 (1968).
Fabri, E.: Nuovo Cimento **11**, 479 (1954).
Fermi, E.: Z. Physik **88**, 161 (1934).
Feynman, R. P., Gell-Mann, M.: Phys. Rev. **109**, 193 (1958).
Fierz, M.: Z. Physik **104**, 553 (1937).
Filthuth, H.: Proceedings of Topical Conference on Weak Interactions, CERN (1969).
Fubini, S., Furlan, G.: Ann. Phys. (N.Y.) **48**, 322 (1968).
Gamow, G., Teller, E.: Phys. Rev. **49**, 895 (1936).
Gasiorowicz, S., Geffen, D. A.: Rev. Mod. Phys. **41**, 531 (1969).
Gell-Mann, M.: Phys. Rev. **92**, 833 (1953).
— Phys. Rev. **111**, 162 (1958).
— Phys. Rev. **125**, 1067 (1962); — Physics **1**, 63 (1964).
— Phys. Letters **8**, 214 (1964).
— *Levy, M.:* Nuovo Cimento **16**, 705 (1960).
— *Ne'eman, Y.:* The Eightfold Way. New York: W. A. Benjamin Publ., 1964.
Gell-Mann, M., Pais, A.: Proceedings of the Glasgow Conference, Pergamon Press (1954).
Gershtein, S. S., Zel'dovich, Y. B.: Soviet Phys. JETP **2**, 576 (1956).
Goldberger, M. L., Treiman, S. B.: Phys. Rev. **110**, 1178 (1958).
Goldhaber, M., Grodzins, L., Sunyar, L.: Phys. Rev. **109**, 1015 (1958).
Goto, T., Imamura, T.: Prog. Theor. Phys. (Kyoto) **14**, 396 (1955).
Klein, O.: Nature **161**, 897 (1948).
Konopinski, E. J., Uhlenbeck, G. E.: Phys. Rev. **60**, 308 (1941).
Landau, L.: Nucl. Phys. **3**, 127 (1957).
Lee, B. W.: Phys. Rev. Letters **12**, 83 (1964).
Lee, T. D., Oehme, R., Yang, C. N.: Phys. Rev. **106**, 340 (1957).
— *Rosenbluth, M., Yang, C. N.:* Phys. Rev. **75**, 905 (1949).
— *Yang, C. N.:* Phys. Rev. **104**, 254 (1956).
— — Phys. Rev. **105**, 1671 (1957).
Lehmann, H., Symanzik, K., Zimmermann, W.: Nuovo Cimento **1**, 425 (1955).
Lüders, G.: Mat. Fys. Medd. Dan. Vid. Selsk. **28**, No. 5 (1954).
Mathur, V. S., Okubo, S., Pandit, L. E.: Phys. Rev. Letters **16**, 371 (1966).

6. — Phys. **1**, 63 (1964).
7. *Cabibbo, N.:* Phys. Rev. Letters **10**, 531 (1963).
8. *Gell-Mann, M., Ne'eman, Y.: The Eightfold way,* pp. 51—51. New York: W. A. Benjamin, Inc. 1964.
9. See for example: *Streater, R. F., Wightman, A. S.: PCT,* spin and statistic and all that. New York: W. A. Benjamin 1964.
10. *Haag, R.:* Phys. Rev. **112**, 669 (1958).
11. *Goldberger, M. L., Treiman, S. B.:* Phys. Rev. **111**, 354 (1958).
12. *Ademollo, M., Gatto, R.:* Phys. Rev. Letters **13**, 264 (1965); — *Fubini, S., Furlan, G.:* Phys. **1**, 229 (1965).
13. *Weinberg, S.:* Phys. Rev. **112**, 1375 (1958).

Volkhard F. Müller
Institut für Theoretische Physik der Universität
D-6900 Heidelberg, Philosophenweg 16
und
Kernforschungszentrum Karlsruhe
D-7500 Karlsruhe

Non Leptonic Decays

Berthold Stech

Contents

1. The Conventional Form of the Non Leptonic Weak Interaction 50
2. Pole Model for p-wave Hyperon Decay 52
3. s-wave Decay Diagramms 54
4. Current Algebra Treatment 56
5. Test of Charged Current Coupling 58
References 59

In this report I do not repeat the basic properties of non leptonic decays for which I refer to last years Summer School report [1] and to the literature. Instead, I intend to give a critical discussion on the theoretical description which is presently employed to describe these decays. It will be seen that part of the success achieved is misleading and important questions are still without answer. The talk concerns itself mainly with hyperon decay processes since they contain more information on our problems than K-decays. We restrict ourselves to the CP conserving part of the interaction.

In Section 1 the interaction responsible for non leptonic decays is presented in this usually assumed form. In Sections 2 and 3 the result of model treatments for p- and s-wave hyperon decays are presented. Section 4 contains the current algebra attempt to relate s- and p-wave amplitudes. Serious inconsistencies show up at this point and are discussed. In Section 5 we present the expected fraction of the decay amplitude in each channel for which violations of the $|\Delta I| = 1/2$ rule should occur if the weak interaction contains charged currents only.

1. The Conventional Form of the Non Leptonic Weak Interaction

The interaction responsible for the non leptonic weak decay processes is assumed to be formed by the product of charged hadron currents. This effective interaction may contain a non local structure – for example due to the virtual exchange of a semi-weakly interacting vector boson.

The transition matrix element for strangeness changing decays reads:

$$\langle f | \mathscr{H}_{\text{eff.}}^{\text{weak}} | i \rangle = \frac{G}{\sqrt{2}} \sin\theta \cos\theta \frac{1}{(2\pi)^4} \int d^4 q\, S^{\mu\nu}(q)$$

$$\cdot \frac{1}{2} \int d^4 x\, e^{iqx} \langle f | T(J_\mu^{1-i2}(x/2) J_\nu^{4+i5}(-x/2) + J_\mu^{4+i5}(x/2) J_\nu^{1-i2}(-x/2)$$

$$+ J_\mu^{4-i5}(x/2) J_\nu^{1+i2}(-x/2) + J_\mu^{1+i2}(x/2) J_\nu^{4-i5}(-x/2)) | i \rangle . \tag{1}$$

We denoted by $S^{\mu\nu}(q)$ a structure tensor (symmetric in μ, ν) which is unknown except for the fact that it has to describe a short range interaction ($r \lesssim 1/M_p$). It is of the form

$$S^{\mu\nu}(q) = g^{\mu\nu} \tag{2}$$

for a point interaction and

$$S^{\mu\nu}(q) = \left(g^{\mu\nu} - \frac{q^\mu q^\nu}{M^2} \right) \frac{M^2}{M^2 - q^2} \tag{3}$$

in the model of an intermediate vector boson of mass M. J_μ^α denotes the normalized V – A combination of $SU\,3 \times SU\,3$ currents. The normalization is fixed by the current commutation rules. The angle θ is the Gell-Mann-Levy-Cabibbo angle. T stands for time ordering. The currents obey the V – A chiral symmetry property

$$[F_5^\alpha(0), J_\mu^\beta(0, \boldsymbol{x})] = [F^\alpha(0), J_\mu^\beta(0, \boldsymbol{x})] \tag{4}$$

where $F^\alpha(0)$ and $F_5^\alpha(0)$ refer to charge and axial charge respectively (at time $t = 0$). In the point interaction limit – if it exist at all – the chiral symmetry of the currents is also valid for the effective Hamiltonian: Decomposing $\mathscr{H}^{\text{weak}}$ into a parity conserving and a parity non conserving part one has

$$[F_5^\alpha(0), \mathscr{H}^{\text{p.c.}}(0)] = [F^\alpha(0), \mathscr{H}^{\text{p.v.}}(0)] , \tag{5a}$$

$$[F_5^\alpha(0), \mathscr{H}^{\text{p.v.}}(0)] = [F^\alpha(0), \mathscr{H}^{\text{p.c.}}(0)] . \tag{5b}$$

According to the Björkén procedure [2] the q-integration in Eq. (1) is likely to diverge quadratically for the structure functions (2) and (3). However, in certain simple models, the quadratically divergent part is a four divergence and thus does not contribute to the transition amplitude of Eq. (1) [3]. The chiral symmetry expressed by Eq. (5) may no more be applicable without corrections if the high q-values play an important role.

$\mathcal{H}^{\text{weak}}$ can be decomposed into an $SU3$ octet and a "27" plet. In spite of the $SU3$ symmetry breaking this decomposition is well defined through the commutation rules as long as the interaction is local. We note that the octet part of the current product is the 6th component of an octet because of the symmetrization in (1) enforced by the assumed CP-invariance.

Experimentally, one observes the $|\Delta I| = 1/2$ selection rule. If nevertheless the form of Eq. (1) is still maintained this result implies small matrix elements of the entire "27" contribution compared to the dominating octet matrix elements.

2. Pole Model for *p*-wave Hyperon Decay

The singularities of the decay amplitudes due to single particle pole terms can be very important and should first be investigated. In *p*-wave hyperon decays (for example Λ-decay) the poles closest to the physical decay point arise from the graphs:

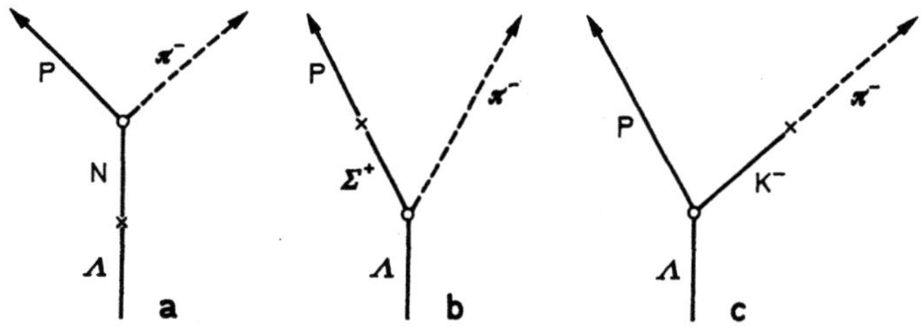

Fig. 1

The weak vertex is denoted by a cross and involves single particle baryon or meson matrix elements of the parity conserving part of the effective Hamiltonian $\mathcal{H}^{\text{p.c.}}$.

It is important to remark here that we can formally modify $\mathcal{H}^{\text{p.c.}}$ by adding a current four divergence

$$\mathcal{H}^{\text{p.c.}}(0) \to \bar{\mathcal{H}}^{\text{p.c.}} = \mathcal{H}^{\text{p.c.}}(0) - \text{const}\, \partial^{\lambda} V_{\lambda}^{7}(0)\,. \tag{6}$$

Such an addition does not change the transition rate because of energy-momentum conservation. Nevertheless it influences the value of one particle matrix elements of $\mathcal{H}^{\text{p.c.}}$ where energy – momentum of initial and final baryon differ. For example, we can fix the constant in Eq. (6) such that

$$\langle \pi | \bar{\mathcal{H}}^{\text{p.c.}}(0) | K \rangle = 0 \tag{7}$$

for, say, momentum transfer $q^2 = 0$. Thus, we can draw the conclusion:

i) It is possible to eliminate the K-pole contribution, thereby modifying other matrix elements of $\mathscr{H}^{\text{p.c.}}$.

If we use a decomposition into $SU3$ invariant and -breaking parts of the vertices appearing in Fig. 1 and recall the mass differences in the propagator we can draw the even more striking conclusion:

ii) The modification (6) leaves unchanged not only the total rate but also the Born part (Fig. (1) itself provided we consider the amplitude to lowest order in $SU3$ breaking only [4]).

Proof: A zero order $SU3$ breaking originating from the divergence of the vector current $\partial^\lambda V_\lambda^7$ in (6) is solely contained in the Born part of the amplitude since only there $SU3$ mass differences appear in the denominator. Consequently, the Born terms for $\bar{\mathscr{H}}^{\text{p.c.}} - \mathscr{H}^{\text{p.c.}}$ must cancel among themselves. *Thus, to this leading order, the meson pole contribution Fig. 1c can entirely be incorporated into the baryon pole diagramms, Figs. 1a, b.* As a further consequence of ii) we note that this leading part of the Born amplitude would *vanish* if for the one particle matrix elements of $\mathscr{H}^{\text{p.c.}}$ the following peculiar relations hold:

$$\begin{aligned} \langle B'|\mathscr{H}^{\text{p.c.}}|B\rangle &= C\,\langle B'|\partial^\lambda V_\lambda^7|B\rangle\,, \\ \langle M'|\mathscr{H}^{\text{p.c.}}|M\rangle &= C\,\langle M'|\partial^\lambda V_\lambda^7|M\rangle \end{aligned} \tag{8}$$

where B and M denote the octet baryon and octet pseudo-scalar meson states, respectively. We will later come back to these statements.

It is now of interest to see to what extent the pole diagramms can describe the p-wave amplitudes. To this purpose we performed an analysis retaining the Born terms only and using the "gauge" [Eq. (7)] where the meson pole disappears. We used $SU3$ symmetry for the strong meson baryon coupling and for the weak vertex for which we incorporate the $|\Delta I| = 1/2$ rule by taking $\mathscr{H}^{\text{p.c.}}$ to transform as an octet. (See end of Section 1; a more general analysis leads to a very small coefficient for the "27" part.) Since $SU3$ breaking through mass terms may be important one should discuss different coupling types which – for non equal baryon masses – give different results.

a) Feynman amplitudes with pseudoscalar coupling, the coupling constants g_{ij} transforming according to $SU3$ with D and F type coupling.

b) Feynman amplitudes with pseudovector coupling, the coupling constants f_{ij} transforming according to $SU3$.

c) Feynman amplitudes with pseudo vector coupling $f_{ij} = \dfrac{g_{ij}}{M_i + M_j}$ with g_{ij} transforming according to $SU3$. The cases a) and c) are not identical since M_i and M_j do refer to the masses of particles at the strong vertex.

The best fit with the three parameters D/F for the strong vertex – (keeping the pion nucleon coupling constant fixed) – and d and f coupling constants for the weak vertex is surprisingly good in all three cases. It shows only small mean square deviations from the experimental numbers.

The pole formulas seem to exhibit the essential features of p-wave decays. In fact, a major change of one of the experimental numbers – for example a sign change – could not be tolerated, it leads to "best fits" with very large mean square deviations from the input data.

The result for the strong D/F ratio is also very satisfactory, one obtains $D/F \simeq +1.5$–1.6 in all three cases a), b), c) in agreement with the analysis in hyperon β-decays [5]. The d and f coupling constants of the weak vertex, although of different magnitude in the three cases, also turn out to have an almost common ratio of about $d/f \simeq -0.9$. Case c) has the smallest mean square error: all fitted amplitudes equal the experimental values within the experimental uncertainties. It gives

$$\begin{aligned} d_{\text{pole}} &\simeq \mp 3.5 \times 10^{-8}\,\text{GeV}\,, \\ f_{\text{pole}} &\simeq \pm 4.1 \times 10^{-8}\,\text{GeV} \end{aligned} \tag{9}$$

where, of course, only the relative sign is determined. We should remember that these numbers refer to the baryon matrix elements of $\bar{\mathscr{H}}^{\text{p.c.}}$ (Eqs. (6) and (7)) from a pole model description of p-wave decays. An attempt to take the K-pole into account would give an ambiguous (in the $SU3$ limit) result because of the freedom expressed by Eq. (6).

3. *s*-wave Decay Diagramms

The s-wave decays occur through the parity changing part of the Hamiltonian. The baryon poles closest to the physical decay point are again those of Figs. 1a and 1b; the $\pi - \text{K}$-vertex (Fig. 1c) vanishes. Contrary to the case of p-wave amplitudes the weak vertex

$$\langle B' | \mathscr{H}^{\text{p.v.}} | B \rangle \tag{10}$$

is forbidden in the strict $SU3$ symmetry limit if crossing and CP-invariance is taken into account. It is, however, not certain whether this selection rule should be taken serious. A similar treatment would then forbid the decay $\text{K}_1^0 \to 2\pi$, which is by no means small. The operator $\mathscr{H}^{\text{p.v.}}$ has the same quantum numbers as the K_1^0 particle except for the fact that the K_1^0 belongs to the 7th component of an octet while $\mathscr{H}^{\text{p.v.}}$ contains an octet part which is a 6th component. Since $SU3$ breaking effects can give a sizeable $\langle 0 | \mathscr{H}^{\text{p.v.}} | \text{K}_1^0 \rangle$ matrix element baryon pole terms can contribute, via the effect of K-dominance, to the baryon matrix elements of $\mathscr{H}^{\text{p.v.}}$.

On the other hand, if indeed K-dominance is the main reason for the baryon s-wave pole terms we have to be aware that the relevant effects can be shifted away to other matrix elements by a redefinition similar to the p-wave case to eliminate the K-pole. We define

$$\hat{\mathscr{H}}^{\text{p.v.}}(0) = \mathscr{H}^{\text{p.v.}}(0) - \frac{\langle \partial | \mathscr{H}^{\text{p.v.}} | \mathrm{K}_1^0 \rangle}{\langle 0 | \partial^\lambda A_\lambda^7 | \mathrm{K}_1^0 \rangle} \cdot \partial^\lambda A_\lambda^7(0) \tag{11}$$

such that

$$\langle 0 | \hat{\mathscr{H}}^{\text{p.v.}} | \mathrm{K}_1^0 \rangle = 0 \tag{12}$$

holds. The new effective parity changing Hamiltonian, although giving the same transition probability, shows no K-dominance any more and the baryon pole contribution is thus drastically diminished at the cost of other matrix elements.

Let us now consider the K* exchange graph of Fig. 2 discussed by *Lee* and *Swift* and *Sakurai* [6].

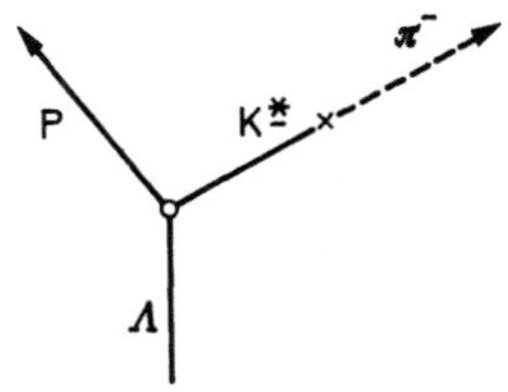

Fig. 2

This Feynman graph gives a contribution which is strongly influenced by the additional term in Eq. (11) and can profit from it. It contains no pole since its absorptive part is zero. Interestingly, this graph provides us with a remarkable model for s-wave decays. Taking the strangeness changing vector current as the interpolating field for the spin 1 K*-meson the amplitude from Fig. 2 is proportional to the divergence of this vector current between the one particle states of initial and final baryons. We have, for example, for the decay $\Lambda \to \mathrm{p} + \pi^-$

$$A_{\Lambda \to p + \pi^-}(\bar{u}_{\mathrm{p}} u_\Lambda) = \text{const} \langle p | \partial^\lambda V_\lambda^{4+i5} | \Lambda \rangle . \tag{13}$$

The model describes parameter-free all s-wave hyperon decay ratios correctly, if we take pure $SU3$ f-type coupling of the nucleon to the vector current. This means octet behaviour for the matrix elements of the divergence in (13) with $d/f \simeq -1/3$ the d/f ratio of the Gell-Mann-Okubo mass formula [7]. In particular, the model gives $A(\Sigma^+ \to \mathrm{N} + \pi^+) = 0$ in agreement with the experimental findings of a very small amplitude in this decay channel. Furthermore, this simple model allows for a prediction of the $K^0 \to 2\pi$ decay rate in terms of the hyperon decay amplitude by simply replacing Λ and proton in (13) by the K and π particle states. This procedure leads to a K-decay amplitude of the correct order of magnitude.

4. Current Algebra Treatment

The pole model for *p*-wave decay and the K* exchange model for *s*-wave decay, although describing the data very well, tell us nothing about the specific form of the weak interaction. Any interaction operator behaving as an octet, whether of the current times current type or not, could be responsable for these decays. The current algebra approach intends now to connect the results of *s*- and *p*-wave decays by using chiral symmetry [Eq. (5)] which is a special property of the current interaction [Eq. (1)].

From the commutation rule (5 b) one obtains in the limit of a vanishing π-meson mass the low energy theorem for the *s*-wave amplitude A [8]

$$A_{\text{soft limit}}(\bar{u}_{B_f} u_{B_i}) = \frac{1}{i f_\pi} \langle B_f | [F^\pi, \mathscr{H}^{\text{p.c.}}] | B_i \rangle . \tag{14}$$

As it is usual here, the *s*-wave poles have been neglected. A similar expression holds for the 2π-decay mode of K^0-decay. Eq. (14) predicts a vanishing amplitude for the decay $\Sigma^+ \to N + \pi^+$ a result the correctness of which is already familiar to us. Taking Eq. (14) for the actual decay amplitude the experimental numbers can be used to obtain the one particle matrix elements of $\mathscr{H}^{\text{p.c.}}$. Within the error limits $\mathscr{H}^{\text{p.c.}}$ behaves in these matrix elements as an octet with d and f coupling:

$$\begin{aligned} d_{\text{s.l.}} &\simeq \mp 1.0 \times 10^{-8}\ \text{GeV}, \\ f_{\text{s.l.}} &\simeq \pm 2.8 \times 10^{-8}\ \text{GeV}. \end{aligned} \tag{15}$$

Note that the sign of $d_{\text{s.l.}}$ is again not determined and may differ from the sign of d_{pole} in Eq. (9). The d/f ratio, according to these numbers, is $\simeq -1/3$, the same as the d/f ratio of the Gell-Mann-Okubo mass formula. This, of course, is not unexpected since the K*-exchange model discussed above (which fitted the data so well) implies the amplitude to be proportional to the divergence of a vector current [Eq. (13)]. This fact, however, brings us now into a peculiar dilemma: the current algebra result provides an expression which, after adjusting d and f, fits the data. However, in this analysis $\mathscr{H}^{\text{p.c.}}$ then behaves as the divergence of a vector current, i.e. the one particle transition matrix elements of both operators have accuratily the same d/f (in the baryon case) and roughly the same baryon to meson ratio. This is a disastrous result for the *p*-wave model. There we saw that, if the one particle matrix elements behave as a divergence as in Eq. (8), the main part of the Born amplitude, which we used to fit the p-wave data, vanishes. In this case the pole model is unreliable and so is a simultaneous fit of s- and p-wave decays as attempted in the literature [9]. The discrepancy is also seen in the comparison of the numbers displayed in Eqs. (9) and (15), in particular in the d/f ratio. It remains even if we try to attribute the difference between f_{pole}, d_{pole} and $f_{\text{s.l.}}$ and $d_{\text{s.l.}}$ to the K-pole (Fig. 1c) which is implicitly contained in f_{pole}, d_{pole}.

We conclude that either the pole formula description of p-wave decay or the use of the uncorrected current algebra result (14) for s-wave decay is incorrect. Clearly, the p-wave pole model should not be taken too serious. However, the proportionality of $\mathscr{H}^{\text{p.c.}}$ to a divergence which we encountered in the current algebra treatment is even more suspicious. The error may be due to the commutation relation (5), but more likely it lies in the neglect of the dependence of the amplitude on the mass of the π-meson and in the neglect of baryon and excited baryon pole contributions to the s-wave. The K*-exchange graph is a good example of a contribution which vanishes in the limit of zero meson mass!

The corrections to the soft limit are most easily seen when using the method of direct insertion of intermediate states into the relevant commutation relation [10, 1]. No good arguments for the smallness of the two types of corrections can be given in the case of a commutator involving $\mathscr{H}^{\text{p.c.}}$. A possible correction was considered in Ref. [1], making use of the removal of K-dominance [Eq. (11)]. More corrections remain to be considered. In this respect it is important to point out that the $\text{K}-\pi$ matrix element of $\mathscr{H}^{\text{p.c.}}$ obtained from the uncorrected soft limit formula is also in violent disagreement (by a factor 4 or 5) with pole determinations of this same matrix element in the $\text{K}\to 3\pi$ decay process as discussed by *Hertel* [11] and *Sutherland* [12] as well as in $\text{K}\to 2\gamma$ decay [13].

Before concluding this section we may consult the simple non-relativistic quark model on the question of the d/f ratio of baryon matrix elements of $\mathscr{H}^{\text{p.c.}}$*. Taking as the effective Hamiltonian in the current current picture the expression

$$\mathscr{H}^{\text{p.c.}}_{\text{eff.}} = \frac{G}{\sqrt{2}} \sin\theta\cos\theta \{:(\psi_n^* \psi_p)(\psi_p^* \psi_\lambda): - :(\psi_n^* \boldsymbol{\sigma} \psi_p)(\psi_p^* \boldsymbol{\sigma} \psi_\lambda): + \text{h.c.}\} \quad (16)$$

it is easy to check that for octet baryon matrix elements $d/f = -1$. The double dots stand for the normal ordering of the non relativistic quark fields. A one-particle operator of the form $\psi_n^* \psi_\lambda$ is proportional to a four divergence and does not contribute to transition matrix

* I am indebted to Dr. *J. Körner* for this suggestion and relevant discussions.

elements. In off-shell treatments it could be eliminated by the modification (6). The value $d/f = -1$ is close to the value obtained from the p-wave pole determination $d/f \simeq -0.9$ [Eq. (9)] and in disagreement with the value $d/f \simeq -1/3$ obtained from the current algebra determination. It is also remarkable that the quark model operator $\mathscr{H}_{\text{eff.}}^{\text{p.c.}}$ does not connect octet with decuplet states.

5. Test of Charged Current Coupling

As mentioned before the success of the s- and p-wave models discussed so far does not give a clue as to the correctness of the basic ansatz of interacting charged currents displayed in Eq. 1. There is, however, a very definite and calculable contribution to the transition amplitudes which is sensitive to the form of the Hamiltonian. This contribution violates the $|\Delta I| = 1/2$ rule completely and thus is worthwhile to be discussed. I refer to the disconnected contribution P arising from the nucleon intermediate state in the current product:

$$P = \frac{G}{\sqrt{2}} \sin\theta \cos\theta \langle \pi^- | J_\mu^{1-i2}(0) | 0 \rangle \langle B_f | J_\nu^{4+i5}(0) | B_i \rangle g^{\mu\nu}. \quad (17)$$

For simplicity the point interaction limit (2) has been taken. The expression (17) describes that part of the decay amplitude in which one particle "travels freely" through one of the current operators. In $\Delta S = +1$ transitions this part clearly exists only in case the final meson is negatively charged. It is an outstanding contribution in this respect and should be one of the relevant terms for breaking the $|\Delta I| = 1/2$ rule. The absolute magnitude of P is easily computed and given in Table 1 in percent of the observed amplitude:

Table 1

	$\Lambda \to p + \pi^-$	$\Sigma^- \to n + \pi^-$	$\Xi^- \to \Lambda + \pi^-$	$K^- \to \pi^- + \pi^0$
s-wave	15%	15%	13%	200%
p-wave	19%	≈150%	12%	—

This result is rather interesting. The contribution which definitely violates the $|\Delta I| = 1/2$ rule is too large by about a factor 2 in amplitude to be compatible with experiment! Although the contribution we calculated is only one of an infinite set it is a significant one since it contains no integration over intermediate states. The numbers show that we have only two alternatives:

a) The hypotheses of charged current coupling is incorrect and neutral currents exist.

b) Non leptonic interaction shows, besides octet dominance (main part of amplitude), an additional suppression or cancellation effect in prominent parts of the weak interaction which violate the $|\Delta I| = 1/2$ rule. (This suppression would have to be a factor 10 in theories without the $\sin\theta$ term!)

If the non dominating part of the weak interaction is suppressed by a factor 2 as $K^{\pm}$-decay indicates the violation of the $|\Delta I| = 1/2$ rule in hyperon, especially Λ-decay, should still be around 7%! Precise experiments are needed.

Connected intermediate states are difficult to handle. The results of calculations of the electromagnetic mass differences are a warning against attempts of this kind. Nevertheless it is worth noting that the estimates lead to the observed octet dominance [14].

The author acknowledges the useful collaboration of Dr. *P. Hertel* and *K. Kemmler* in part of this work and their help in numerical calculations.

References

1. *Stech, B.:* Springer Tracts in Modern Physics **50**, 84 (1969).
2. *Björkén, J. D.:* Phys. Rev. **148**, 1467 (1966).
3. *Bouchiat, C., Iliopoulos, J., Prentki, J.:* Nuovo Cimento **56**, 1150 (1968).
4. *Adler, S. L., Dashen, R.:* Current algebra and applications to particle physics, Benjamin N.Y., 1968.
5. *Brene, N., Roos, M., Sirlin, A.:* Nucl. Phys. B **6**, 255 (1968).
6. *Lee, B. W., Swift, A. R.:* Phys. Rev. **136**, B 229 (1964). — *Sakurai, J. J.:* Phys. Rev. **156**, 1508 (1967).
7. *Okubo, S.:* Prog. Theoret. Phys. (Kyoto) **27**, 949 (1962).
8. *Sugawara, H.:* Phys. Rev. Letters **15**, 870, 997 (E) (1965). — *Suzuki, M.:* Phys. Rev. Letters **15**, 986 (1965).
9. *Hara, Y., Nambu, Y., Schecter, J.:* Phys. Rev. Letters **16**, 380 (1966). — *Brown, L. S., Sommerfield, C.:* Phys. Rev. Letters **16**, 751 (1966).
10. This method is equivalent to the q_0-dispersion theoretical procedure developed by *Fubini, S., Furlan, G.:* Ann. Phys. (N.Y.) **48**, 322 (1968). — *de Alfaro, V., Rosetti, C.:* Invited Report at the 1967 Conference of the Italian Physical Society.
    See *Stech, B.:* Lectures on weak decays of hadrons, NORDITA, Copenhagen 1968. – *Dahmen, H. D., Rothe, K. D., Schülke, L.:* Nucl. Phys. B **7**, 472 (1968). – *Stech, B.:* Proceedings of the topical conference on weak interactions. CERN 1969, 271.
11. *Hertel, P.:* Z. Physik **226**, 7 (1969).
12. *Sutherland, D. B.:* Nucl. Phys. B (to be published).
13. *Hertel, P.:* Heidelberg preprint (to be published).
14. *Hara, Y.:* Progr. Theoret. Phys. (Kyoto) **37**, 710 (1967).

Professor Dr. *Berthold Stech*
Institut für Theoretische Physik
und Mechanik der Universität
D-6900 Heidelberg, Philosophenweg 16

Current Algebra and Weak Interactions

B. RENNER

Contents

Summary 60
1. The Basic Principles of Current Algebra 60
2. Sum Rule Tests 64
3. Low Energy Theorems 67
4. K_{13} Decay 69
5. K_{14} Decay 72
6. The Structure of Low Energy Theorems 74
7. The Open Questions 77
References 78

Summary

These are the notes of three lectures given at the Summer Institute in Theoretical Physics at the University of Karlsruhe, July 1969. The basic principles of current algebra are discussed, and their application to weak interactions. No claim is made to give an exhaustive survey, and certain topics have been deliberately omitted with reference to other lectures. Many of the preliminary concepts to this discussion have been explained in further detail in the lectures of *S. Gasiorowicz*.

1. The Basic Principles of Current Algebra

The basic objects of current algebra are two octets of hermitean vector currents V_μ^i $(i=1 \dots 8)$ and axial vector currents A_μ^i $(i=1 \dots 8)$. They are identified through the hadron electromagnetic current

$$J_\mu^{\mathrm{el}} = V_\mu^3 + \frac{1}{\sqrt{3}} V_\mu^8 \tag{1}$$

and the hadron weak current

$$J_\mu^{\mathrm{W}} = \cos\theta (V+A)_\mu^{(1+i2)} + \sin\theta (V+A)_\mu^{(4+i5)}, \tag{2}$$

which together with the lepton weak current l_μ, constitutes the weak interaction energy density

$$H^{\mathrm{W}} = \frac{G_F}{\sqrt{2}} \cdot \frac{1}{2} \{(J_\mu^{\mathrm{W}} + l_\mu)(J^{\mathrm{W}+\mu} + l^{+\mu}) + (J_\mu^{\mathrm{W}+} + l_\mu^{+})(J^{\mathrm{W}\mu} + l^\mu)\} . \quad (3)$$

The following equal-time commutators have been proposed by *Gell-Mann* [1] between the currents and their charges $Q = \int J_0(x)\,(\mathrm{d}x)^3$

$$[Q_V^i, V_\mu^j] = i f^{ijk} V_\mu^k , \quad (4)$$

$$[Q_V^i, A_\mu^j] = i f^{ijk} A_\mu^k , \quad (5)$$

$$[Q_A^i, V_\mu^j] = i f^{ijk} A_\mu^k , \quad (6)$$

$$[Q_A^i, A_\mu^j] = i f^{ijk} V_\mu^k . \quad (7)$$

The algebraic structure of these commutators is $SU_3 \times SU_3$. We wish to expose four physical principles, which are contained in the current algebra postulates, Eqs. (4)–(7).

(1) The quantum numbers of weak and electromagnetic currents are specified through Eqs. (1), (2), and (4). On the level of isospin (I) symmetry, Eq. (4) specifies $I = 1$ for the strangeness conserving weak current and $I = 1/2$ and $\Delta S/\Delta Q = 1$ for the strangeness changing weak current. Weak transitions with $\Delta S = 2$ are excluded. There is no positive evidence for the existence of weak currents with other quantum numbers, but negative evidence is sometimes hard to obtain. Only in semileptonic decays are the individual currents isolated; conclusions from non-leptonic decays are always fraught with additional hypotheses.

(1a) $\Delta S = 0$: Weak currents with $I \geqq 2$ can only operate on targets with $I \geqq 1$. In pion β-decay an $I = 2$ weak current can, by generalized Bose-statistics*, only have matrix elements

$$\langle \pi | J_\mu^{(I=2)} | \pi' \rangle = F_2(\Delta^2)\,(p - p')_\mu , \quad (8)$$

whose effects would be hard to detect. In nuclear β-decay, currents of $I \geqq 2$ can only show up in corrections to the usual impulse approximation, which describes the nuclear β-transition as a single-nucleon β-transition inside the nucleus. The size of these corrections is estimated to be about ten per cent [2].

(1b) $\Delta S = 1$: Deviations from $I(J) = 1/2$ could be seen in Ξ-decay (not feasible) and in comparing K^+ with K^0 decay (the data are consistent with $I(J) = 1/2$, but the analysis is complicated by $(K^0, \bar{K}^0)$ mixing). $\Delta S/\Delta Q = -1$ currents could be seen in $\Sigma^+ \to n\,e^+\,\nu$ (evidence questionable),

* To see this, consider $\langle 0 | J_\mu^{(2)} | \pi' \pi \rangle$, where the pions have a spatially symmetric wave function. The matrix element has to reflect this symmetry.

$\Xi^0 \rightarrow \Sigma^- e^+ \nu$ (not feasible), $K^0 \rightarrow \pi^+ e^- \bar{\nu}$ and $\bar{K}^0 \rightarrow \pi^- e^+ \nu$ [the data are consistent with the absence of these transitions, but the analysis is complicated by $(K^0, \bar{K}^0)$ mixing].

(1c) $\Delta S = 2$: In principle such currents could be seen in leptonic Ξ and Ω^- decays, this is not easy to check experimentally.

In summary: though the present data are consistent with the current algebra assignments of quantum numbers, there are still gaps in their confirmation, which can only be filled by data from neutrino reactions.

The validity of current commutation relations also constrains the behaviour of the currents under charge conjugation and time reversal [3]. Certain types of second class currents are excluded.

(2) Current algebra contains the CVC theory, in which the isovector part of the electromagnetic current and the strangeness conserving part of the weak vector current are assigned to the same isospin triplet of currents. Consequences:

(a) The strangeness conserving weak vector current is conserved; its scale is fixed in terms of the target isospin:

$$\text{Example: pion } \beta\text{-decay: } \langle \pi^0 | J_\mu^W | \pi^\pm \rangle = \sqrt{2} \cdot \cos\theta \cdot (p^0 + p^\pm)_\mu . \quad (9)$$

(b) The form factors of the weak current are related to the form factors of the isovector electromagnetic current:

Example: weak magnetism:

$$\langle p | J_\mu^W | n \rangle = \bar{u}_p [\gamma_\mu + (\mu_p - \mu_n) \sigma_{\mu\nu} (p^n - p^p)^\nu] u_n + \text{axial part} . \quad (10)$$

Let us digress at this point and ask to what extent the CVC-theory conforms with a simple dynamical framework.

In the approximation of vector-meson dominance, the matrix elements of any two currents with the same quantum numbers are given by

$$\langle f | V_\mu | i \rangle \approx \frac{\langle f \varrho | i \rangle \langle \varrho | V_\mu | 0 \rangle}{(p^f - p^i)^2 - m_\varrho^2} . \quad (11)$$

Imagine that the weak vector current and the isotriplet partner of the electromagnetic current are not identical. In the domain of ϱ-dominance, their matrix elements nevertheless have to be proportional, with a universal factor originating from $\langle \varrho | V_\mu | 0 \rangle$.

Corrections to the vector meson pole-model can be introduced by taking into account the $(\pi\pi)$ continuum. In linearized unitarity relations, the discontinuity of the form factors is determined:

$$(2i)\,\text{Im}\, \langle f | J_\mu | i \rangle = \sum_n \langle 0 | J_\mu | n \rangle \langle n | \bar{f} i \rangle^x , \quad (12)$$

in terms of the strong interaction matrix element $\langle n|\bar{f}i\rangle^x$ and the pion form factor, disregarding all other intermediate channels except the $(\pi\pi)$ continuum with the ϱ-resonance. For the pion form factor, we use again a linearized unitarity relation; we find it proportional to $\exp(i\delta(\Delta^2))$ [$\delta(\Delta^2)$: pion p-wave phase shift] in terms of an Omnès function

$$F(\Delta^2)=P(\Delta^2)\exp\left\{\frac{\Delta^2}{\pi}\int\limits_{4m_\pi^2}^{\infty}\frac{\delta(t)\,\mathrm{d}t}{(t-\Delta^2)\cdot t}\right\}. \tag{13}$$

$P(\Delta^2)$ is a subtraction polynomial, assumed to be constant in simple dynamical models.

So far there is no scope for the matrix elements of the weak vector current and those of the isotriplet partner of the electromagnetic current to differ by more than a scale factor. Nontrivial differences do not arise until the effects of several continuum channels are taken into account.

To confirm the CVC-theory in its fundamental significance, tests would be needed of a higher accuracy than that of the vector meson or $(\pi\pi)$ channel dominance approximation. Weak magnetism can be explained to this approximation by quasi-dynamical arguments, but the scale of the currents has to be postulated. We will find the same situation in trying to explain the chiral aspect of current algebra.

(3) Current algebra demands exact SU_3 commutation relations, despite the fact that SU_3 is not an exact symmetry of strong interactions. In the SU_3 symmetry limit, the current algebra relations (4) and (5) are equivalent to the Cabibbo theory of semileptonic decays [3]. In broken SU_3, current algebra retains (in fact defines) octet behaviour of currents; the states need no longer form pure SU_3 multiplets. The success of the Cabbibbo theory in describing physical decays can only in part be credited to current algebra; it is mainly due to approximate SU_3 behaviour of the hadron states.

Current algebra is only confirmed in its proper significance by demonstrating the validity of the commutators even under SU_3 violations, for instance by connecting otherwise uncorrelated SU_3 violations. (See the discussion of K_{13} decay.)

(4) We have not yet discussed the significance of commutators (6) and (7). In any sufficiently regular theory, Eqs. (5) and (6) are equivalent up to chargeless currents of magnetic type whose time components vanish upon spatial integration. Eq. (7) completes the statement of weak interaction universality, by specifying the scale of the axial current. As a consequence of Eq. (7), the hadron weak charge Q^W and its hermitean adjoint Q^{W+} form an SU_2 algebra together with their commutator, as

do the lepton weak charge and its hermitean adjoint (to be verified by explicit canonical commutation of the lepton fields).

The requirement of a weak SU_2 algebra is sufficient to determine the commutators in Eq. (7) with (4), (5), and (6) assumed. Its origin is yet completely unclear.

We will investigate to what extent the results of commutators (7) can be argued for on other grounds.

2. Sum Rule Tests

Ideally, tests of current algebra should be simple: just insert a complete set of intermediate states into a commutator and obtain a sum rule for the matrix elements of the currents. The practical limitation to its use is the inaccurate knowledge of the many-particle contributions at high energies, due to limited experience in neutrino reactions. The problem is now how to rewrite the sum rules so as to minimize reference to unknown data. Consider for example the commutator of weak charges between protons*

$$\sum_n \left\{ \left\langle p \left| Q^{\mathrm{W}} \frac{|n\rangle\langle n|}{N_{\mathrm{n}}} Q^{\mathrm{W}+} \right| p \right\rangle - \left\langle p \left| Q^{\mathrm{W}+} \frac{|n\rangle\langle n|}{N_{\mathrm{n}}} Q^{\mathrm{W}} \right| p \right\rangle \right\} = (2\cos^2\theta + 4\sin^2\theta)\, N_{\mathrm{p}}\,. \tag{14}$$

Through the space-integration in the charges, the momentum of the intermediate states is selected to be equal to the momentum of the external states. This allows us to rewrite the matrix elements of the charges in terms of current divergences

$$\sum_n \delta^3(p_{\mathrm{n}} - p_{\mathrm{p}}) \frac{|\langle p|\partial J^{\mathrm{W}}|n\rangle|^2 - |\langle p|\partial J^{\mathrm{W}+}|n\rangle|^2}{4E_{\mathrm{p}}E_{\mathrm{n}}(E_{\mathrm{p}} - E_{\mathrm{n}})^2} = (2\cos^2\theta + 4\sin^2\theta)\,. \tag{15}$$

Degenerate intermediate states should be given a separate treatment. The result is the same, if a small mass difference is introduced**.

Adler [4] observed that the divergence of the weak current is measured in ν-reactions:

$$\sum_{\substack{\text{lepton}\\ \text{polarization}}} |\langle n, l|T|p, \nu\rangle|^2 = 4G_F^2 \langle p|J_\lambda^{\mathrm{W}+}|n\rangle \langle n|J_\mu^{\mathrm{W}}|p\rangle \left(p_\mu^l p_\lambda^\nu + p_\mu^\nu p_\lambda^l - g_{\mu\lambda}(p^\nu p^l)\right). \tag{16}$$

* N_{n} stands for the state normalization. We choose: $\langle p|p'\rangle = 2E \cdot (2\pi)^3\, \delta^3(p - p')$.

** Consider that $\dfrac{\langle p|\partial J|n\rangle}{i(E_{\mathrm{p}} - E_{\mathrm{n}})} = \langle p|J_0|n\rangle$ which is not sensitive to small mass differences.

In the case of forward scattering, neglecting the lepton mass, this equals

$$\Sigma = 8\, G_F^2 |\langle n|\partial J^{\mathrm{W}}|p\rangle|^2 \frac{E^\nu E^l}{(E^\nu - E^l)^2}\,. \tag{17}$$

For $E^\nu > (m_{\mathrm{n}}^2 - m_{\mathrm{p}}^2)/2m_{\mathrm{p}}$, Σ can be expressed in terms of a forward differential cross section

$$\frac{\mathrm{d}\sigma}{\mathrm{d}\Omega_l}(\nu p \to l n) = G_F^2 \frac{|\langle n|\partial J^{\mathrm{W}}|p\rangle|^2 (m_{\mathrm{p}}^2 + 2 m_{\mathrm{p}} E^\nu - m_{\mathrm{n}}^2)}{8\pi^2 \cdot m_{\mathrm{p}}^2 \cdot (m_{\mathrm{n}}^2 - m_{\mathrm{p}}^2)^2}\,. \tag{18}$$

To insert Eq. (18) into the sum rule (15) we have to match the selection of the momentum transfer between external and intermediate states. In (18), the momentum transfer vanishes:

$$(p^{\mathrm{n}} - p^{\mathrm{p}})^2 = (p^\nu - p^l)^2 = 0 \tag{19}$$

in (15) it is determined by the common momentum q of the external and intermediate states:

$$(p^{\mathrm{n}} - p^{\mathrm{p}})^2 = (E^{\mathrm{n}} - E^{\mathrm{p}})^2 \approx \frac{1}{4q^2}(m_{\mathrm{n}}^2 - m_{\mathrm{p}}^2) \quad \text{for large } q\,. \tag{20}$$

So we have to take $q \to \infty$ to write* in (15)

$$G_F^2(2\cos^2\theta + 4\sin^2\theta)$$

$$= \int\limits_{m_N}^{\infty} \mathrm{d}W \frac{8\pi^2 m_{\mathrm{p}}^2}{(m_{\mathrm{p}}^2 + 2m_{\mathrm{p}}E - W^2)^2}\left[\frac{\mathrm{d}^2\sigma^{\bar\nu}(E,W)}{\mathrm{d}\Omega_l\,\mathrm{d}W} - \frac{\mathrm{d}^2\sigma^{\nu}(E,W)}{\mathrm{d}\Omega_l\,\mathrm{d}W}\right], \tag{21}$$

E: neutrino (antineutrino) energy,
W: effective mass of the hadron system produced.

This sum rule is not yet testable with present data; however it has an interesting consequence as $E \to \infty$.

If we make the assumption that the main contributions to the total production cross section are given by hadron states of small invariant mass W, we can, for sufficiently large E, neglect the W-dependence of the kinematic factor in the integrand of (21)** and derive

$$G_F^2(\cos^2\theta + 2\sin 2\theta) = \lim_{E\to\infty}\left(\frac{\pi}{E}\right)^2 \left|\frac{\mathrm{d}\sigma^{\bar\nu}(E)}{\mathrm{d}\Omega_l} - \frac{\mathrm{d}\sigma^{\nu}(E)}{\mathrm{d}\Omega_l}\right| \tag{22}$$

* The limit $q \to \infty$ is a technical device to select a suitable sum rule. In Eq. (21), the differential cross section refers to a stationary target.

** The vanishing of the denominator in (21) is compensated by a threshold factor in the differential cross section.

in the forward direction. *Adler* [5] has generalized this relation to non-forward directions, under additional assumptions:

$$\lim_{E\to\infty}\left(\frac{d\sigma^{\bar\nu}}{dt}-\frac{d\sigma^{\nu}}{dt}\right)=\frac{G_F^2}{\pi}(\cos^2\theta+2\sin^2\theta)\quad\text{(independent of } t\text{)}. \tag{23}$$

Since the accessible range of the momentum transfer t rises linearly with E, Eq. (23) implies linearly rising total cross-sections, as expected from lowest-order Fermi coupling. This makes the limit inaccessible, since at large E one expects to see departures from first-order weak interactions. By approximate saturation of the forward scattering sum rule (21), *Adler* and *Gilman* [6] have concluded that with a neutrino energy of about 5 GeV the sum rule may be testable to within ten per cent accuracy.

According to recent data [7], the total neutrino cross-section is still linearly rising with energy in this domain. Until neutrino data become available, current algebra sum rules are tested in the meson-pole dominance approximation (PCAC). Matrix elements of the strangeness conserving axial-vector current are approximated by a pion pole term [3]

$$\langle n|\partial A|p\rangle\approx\frac{\langle n|T|p\pi\rangle\langle\pi|\partial A|0\rangle}{(p_p-p_n)^2-m_\pi^2}. \tag{24}$$

The best test for the validity of this approximation is the Goldberger-Treiman relation, which is derived from Eq. (24) with proton and neutron states at $(p_p-p_n)^2\approx 0$

$$\langle n|\partial A|p\rangle=(-i)\,2m_N\cdot g_A\bar u_n\gamma_5 u_p, \tag{25}$$

$$\langle n|T|p\pi\rangle=i(G_{np\pi})\,\bar u_n\gamma_5 u_p=\sqrt{2}\,i\,G_{NN\pi}\bar u_n\gamma_5 u_p, \tag{26}$$

$$\langle\pi|\partial A|0\rangle=\sqrt{2}F_\pi m_\pi^2, \tag{27}$$

substituting into (24):

$$m_N\cdot g_A\approx F_\pi G_{NN\pi}. \tag{28}$$

Eq. (28) is the Goldberger-Treiman relation. It connects the nucleon weak axial vector coupling constant g_A with the weak pion decay constant F_π and the Yukawa coupling constant $G_{NN\pi}$, to an accuracy of about ten per cent. Qualitatively this discrepancy can be understood in terms of a cut correction [8] to Eq. (24).

Separating the strangeness conserving axial vector part in sum rule (15), and making the pion pole dominance approximation (24) for the matrix elements of its divergence, we obtain the Adler-Weisberger relation [9]:

$$1=|g_A|^2+2F_\pi^2\int\limits_{(m_N+m_\pi)^2}^{\infty}\frac{ds}{(s-m_p^2)^2}(\sigma_{\pi^-p}^{tot}(s)-\sigma_{\pi^+p}^{tot}(s))\frac{2p_\pi^L m_p}{\pi}. \quad (29)$$

3. Low Energy Theorems

We compare the Adler-Weisberger relation with a forward dispersion relation in πp scattering. A variable ν is defined by $s=m_p^2+2\nu+m_\pi^2$, we have:

$$T(\nu)=\frac{1}{2}(T^{(\pi^- p)}(\nu)-T^{(\pi^+ p)}(\nu)), \quad (30)$$

$$\begin{aligned}T(\nu)=\frac{2\nu}{\pi}\int\limits_0^\infty\frac{\operatorname{Im}T(\nu')}{\nu'^2-\nu^2}d\nu'=\frac{4\nu G_{NN\pi}^2[(m_p-m_n)^2-m_\pi^2]}{[m_p^2-m_n^2+m_\pi^2]^2-4\nu^2}\\+\frac{\nu}{\pi}\int\limits_{(m_p\, m_\pi)}^{\infty}\frac{2m_p p_\pi^L(\sigma_{(-)}^{tot}(\nu')-\sigma_{(+)}^{tot}(\nu'))}{\nu'^2-\nu^2}d\nu',\end{aligned} \quad (31)$$

$$\begin{aligned}F_\pi^2\frac{dT}{d\nu}\bigg|_{\nu=0}=F_\pi^2\frac{4G_{NN\pi}^2[(m_p-m_n)^2-m_\pi^2]}{[m_p^2-m_n^2+m_\pi^2]^2}\\+\frac{2F_\pi^2}{\pi}\int\limits_{(m_p+m_\pi)^2}^{\infty}\frac{2m_p p_\pi^L(\sigma_{(-)}^{tot}(s)-\sigma_{(+)}^{tot}(s))}{(s-m_p^2-m_\pi^2)^2}ds.\end{aligned} \quad (32)$$

We recognize the similarity of the continuum integrals in the Adler-Weisberger relation (29) and the dispersion relation (32) in the limit $m_\pi\to 0$. In the same limit, keeping $m_p \neq m_n$, the Born term tends to

$$4F_\pi^2 G_{NN\pi}^2/4m_N^2=g_A^2 \quad (33)$$

according to the Goldberger-Treiman relation (28). We have now gained an alternative expression for the Adler-Weisberger relation as a low-energy theorem for an unphysical amplitude:

$$F_\pi^2\frac{d}{d\nu}T(\nu=0,\ p_\pi^2=0)=1. \quad (34)$$

In the following we will describe a direct derivation of the low-energy theorem (34), without the circuit via sum rules, infinite-momentum limit and equivalence to dispersion relations, using instead a more extended concept of PCAC. With the LSZ reduction formalism, we can represent a virtual pion forward scattering amplitude as

$$T(\pi^- p) = i(m_\pi^2 - q^2)^2 \int e^{iqx} \langle p | T\{\phi^+(x), \phi^-(0)\} | p \rangle (dx)^4, \qquad (35)$$

and extend this definition to values of $p_\pi^2 = q^2 \neq m_\pi^2$. $\phi(x)$ is the interpolating pion field, which is yet to be specified. We have the freedom of defining the pion field through the divergence of the axial-vector current

$$\phi^i = \partial A^i / F_\pi m_\pi^2 . \qquad (36)$$

Substituting (35) we perform two partial integrations

$$\begin{aligned} T(\pi^- p) &= i \frac{(m_\pi^2 - q^2)^2}{(F_\pi m_\pi^2)^2} \int e^{iqx} \langle p | T\{\partial A^+(x), \partial A^-(0)\} | p \rangle (dx)^4 \\ &= i \frac{(m_\pi^2 - q^2)^2}{(F_\pi m_\pi^2)^2} \{ q^\mu q^\nu \int e^{iqx} \langle p | T\{A_\mu^+(x), A_\nu^-(0)\} | p \rangle (dx)^4 \\ &\qquad - i q^\nu \langle p | V_\nu^3(0) | p \rangle - i \langle p | \sigma(0) | p \rangle \} , \end{aligned} \qquad (37)$$

where $\sigma(0)$ has been defined through $[Q_A^+, \partial A^-] = i\sigma$. Combining (37) with the analogous equation for $T(\pi^+ p)$, Eq. (34) is reproduced from the vector current term.

The main problem is now that of relating the unphysical low-energy theorem in Eq. (34) to physical pion-nucleon scattering. More generally, one may ask about the relations of the amplitudes in Eqs. (35) and (37), when extrapolated to $q^2 \neq m_\pi^2$, to the physical pion scattering amplitude defined only at $q^2 = m_\pi^2$.

The off-shell amplitude in Eq. (35) is clearly dependent on the precise definition of the pion interpolating field. Every local operator $a(x)$ with $\langle 0 | a(0) | \pi \rangle \neq 0$ can serve to define a pion field

$$\phi_a(x) = \frac{a(x)}{\langle 0 | a(0) | \pi \rangle} . \qquad (38)$$

On the pion mass shell, at $q^2 = m_\pi^2$, every pion field reproduces the same physical pion transition amplitude. This can be understood by considering the fact that at $(q^2 = m_\pi^2)$, the factors $(q^2 = m_\pi^2)$ project out the pole residues of the integral, which factorize [10] as in Eq. (24) into the physical transition amplitude and a normalization factor $\langle 0 | \phi_a | \pi \rangle$, which is equal to unity according to Eq. (38).

Off the mass-shell, at $q^2 = m_\pi^2$, the amplitude depends on the nature of the chosen interpolating pion field. Certain variations with q^2 are dynamically required, for instance the variations associated with the neutron Born pole in Eq. (32). These have to be recognized and separated. Other variations may be a property of special interpolating fields, for instance an extra factor of (q^2/m_π^2) in choosing an interpolating field $\tilde{\phi}_a$

$$\tilde{\phi}_a = -\frac{\square}{m_\pi^2}\phi_a \tag{39}$$

rather than ϕ_a.

We now make an important assumption: with the choice of the axial-vector divergence as a special interpolating pion field (Eq. (36)), only the dynamically interpretable variations with q^2 are regarded as significant in an extrapolation from m_π^2 to zero mass; once these "necessary variations" have been taken account of, the assumption of smooth extrapolation can be made.

This assumption has been very successful in practice. We will give a few examples in the next sections. Obviously one will ask in what respect the axial-vector divergence is distinguished over other forms of pion fields. An answer to this question may be found in the considerations of the chiral symmetry limit [11], which we will discuss later.

4. K_{l3} Decay

As an example of a low-energy theorem we consider the matrix element of the hadron vector current in K_{l3} decay

$$\langle\pi^0(q)|V_\mu^{K^-}|K^+(p)\rangle = -\frac{1}{\sqrt{2}}(F^+(t)(p+q)_\mu + F^-(t)(p-q)_\mu). \tag{40}$$

The normalization has been arranged so that $F^+(0)=1$ in the SU_3-symmetry limit. We extrapolate off the pion mass shell

$$\begin{aligned}\langle\pi^0(q)|V_\mu^{K^-}|K^+(p)\rangle &= i\frac{(m_\pi^2-q^2)}{F_\pi m_\pi^2}\int e^{iqx}\langle 0|T\{\partial A^0(x), V_\mu^{K^-}(0)\}|K^+(p)\rangle(dx)^4\\ &= \frac{(m_\pi^2-q^2)}{F_\pi m_\pi^2}q^\lambda\int e^{iqx}\langle 0|T\{A_\lambda^0(x), V_\mu^{K^-}(0)\}|K^+(p)\rangle(dx)^4\\ &\quad (-i)\frac{(m_\pi^2-q^2)}{F_\pi m_\pi^2}\langle 0|[Q_A^0, V_\mu^{K^-}]|K^+(p)\rangle\,.\end{aligned} \tag{41}$$

In the limit $q\to 0$, the following relation is obtained [12]

$$F^+(m_K^2) + F^-(m_K^2) = (F_K/F_\pi) + O(m_\pi^2)\,. \tag{42}$$

This relation is a good illustration of how SU_3-violations are correlated through current algebra. We have

	$F^+(m_K^2)/F^+(0)$	$F^-(m_K^2)/F^+(0)$	$F_K/F_\pi F^+(0)$	$O(m_\pi^2)$
in SU_3	$1+\left(\frac{m_p}{m_V}\right)^2 \approx 1.25$	0	1	0.25
exp.	$1+\left(\frac{m_K}{m_{K^*}}\right)^2 \approx 1.45$	?	1.25	0.1 ± 0.2

In estimating $F^+(m_K^2)$, we used a K^*-pole model, which is consistent with the measurements. Since the extrapolation distance decreases [m_π^2 as compared with $(m_p^2)_{av.}$], the extrapolation error should decrease, except for special dynamical circumstances like prominence of a low lying κ-resonance or a strong threshold interaction in the $(K\pi)$ system. So we predict:

$$F^-(m_K^2)/F^+(0) \approx -0.1 \pm 0.2 \tag{43}$$

and unless a low-lying κ introduces strong t-variations, we find

$$\xi = F^-(t)/F^+(t) \approx -0.1 \pm 0.2 \tag{44}$$

If the SU_3 violations are not found correlated this way, we would have to give up one of two principles:

(a) Current algebra has an exact SU_3 structure, despite the fact that SU_3 is not an exact symmetry group of strong interactions.

(b) Soft-pion extrapolations are smooth.

Only if there were a low-lying κ resonance or some other strong threshold variation, could we give up (43) and (44) while maintaining the validity of current algebra and of the other low energy theorems.

A similar relation can be derived in the limit of vanishing K-meson momentum

$$F^+(m_\pi^2) - F^-(m_\pi^2) = F_\pi/F_K + O(m_K^2). \tag{45}$$

Given the large extrapolation error, this relation is consistent with (42). Since, however, the significance of these relations is in correlating SU_3 violations, K-pole dominance may not be accurate enough for this purpose. The relative distances of meson poles and correction cuts from the low energy point, which is 1 : 9 in the case of pion-pole dominance and meson-pole dominance in SU_3 symmetry, deteriorates to 1 : 2 for K-pole approximations.

Another soft-pion theorem could be derivable from the vertex

$$\langle \pi^0(q) | \partial V | K^+(p) \rangle = \frac{1}{\sqrt{2}} i((m_K^2 - m_\pi^2) F^+(t) + t F^-(t)) . \qquad (46)$$

Its soft-pion limit is given by

$$\frac{1}{\sqrt{2}} m_K^2 (F^+(m_K^2) + F^-(m_K^2)) = -\frac{1}{F_\pi} \langle 0 | [Q_A^0, \partial V^{K^-}] | K^+(p) \rangle + O(m_\pi^2) . \qquad (47)$$

The commutator is yet unknown. Assumptions have to be made about the algebraic behaviour of current divergences in the $SU_3 \times SU_3$ algebra. It can be seen that current divergences belong to the same operator multiplet as the symmetry breaking part of the energy density:

$$\frac{\mathrm{d}Q}{\mathrm{d}t} = \int \partial J (\mathrm{d}x)^3 = i [\int \mathscr{H}_B (\mathrm{d}x)^3, Q] . \qquad (48)$$

Without any commitment as to the closeness of an $SU_3 \times SU_3$ symmetry limit, we assign $\mathscr{H}_B$ to a simple operator multiplet of $SU_3 \times SU_3$, in analogy with the Gell-Mann Okubo assignment [1]. There are two multiplets without exotic operators (i.e. operators with non-octet quantum numbers)

$(1, 8) + (8, 1)$	$(3, 3^*) + (3^*, 3)$	
$[Q_V^i, h^j] = i f^{ijk} h^k$	$[Q_V^i, u^j] = i f^{ijk} u^k$	
$[Q_V^i, g^j] = i f^{ijk} g^k$	$[Q_V^i, v^j] = i f^{ijk} v^k$	
$[Q_A^i, h^j] = i f^{ijk} g^k$	$[Q_A^i, u^j] = -i d^{ijk} v^k$	(49)
$[Q_A^i, g^j] = i f^{ijk} h^k$	$[Q_A^i, v^j] = i d^{ijk} u^k$	
h^j scalar octet	u^j scalar nonet	
g^j pseudoscalar octet	v^j pseudoscalar nonet.	

The ansatz $\mathscr{H}_B = h^8$ is unsuitable, since it would imply $SU_3 \times SU_3$ symmetry with conserved axial charges in the SU_3 symmetry limit. Furthermore it would imply $\partial A^\pi = 0$.

The simplest consistent assignment is

$$\mathscr{H}_B = u^0 + c u^8 \qquad (50)$$

in the multiplet $(3, 3^*) + (3^*, 3)$. This assignment allows us to take independently the SU_3 and chiral symmetry limit, but we still maintain a strong correlation between vector and axial vector divergences. In this model:

$$\partial V^{\mathrm{K}} = i c u^{\mathrm{K}} \cdot \frac{\sqrt{3}}{2} \tag{51}$$

$$[Q_A^0, \partial V^{\mathrm{K}}] = \frac{3c}{4} \cdot \frac{\partial A^{\mathrm{K}}}{(\sqrt{2} - c/2)} . \tag{52}$$

Substituting (52) into (47), we get consistency with (42) if we set $c \approx -\sqrt{2}$, neglecting terms of order m_π^2. Keeping terms of order m_π^2, the best value for c comes out [13]

$$\left(\frac{c + \sqrt{2}}{\sqrt{2}}\right) \approx \frac{m_\pi^2}{(m_{\mathrm{p}}^2)_{\mathrm{av.}}} ; \quad c \approx -1.25 . \tag{53}$$

If Eq. (50) is correct, the symmetry-breaking constant c would have a universal significance similar to the Cabibbo angle. It is not easily accessible, yet so far all its estimates have been consistent with (53).

5. K_{l4} Decay

We consider only the axial current decay channel

$$\langle \pi^+(p)\, \pi^-(q) | A_\mu^{\mathrm{K}^-} | K^+(k) \rangle = i\{F_1(p+q)_\mu + F_2(p-q)_\mu + F_3(k-q-p)_\mu\} . \tag{54}$$

In the low-energy domain, there is no reason known for strong variations of F_1 and F_2; however F_3 contains a K-pole, according to the diagram below.

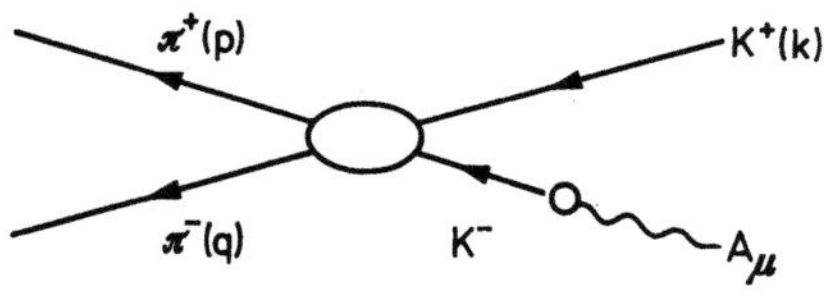

Fig. 1

We make the ansatz

$$F_3 = \frac{\sqrt{2}\, F_{\mathrm{K}}\, T(\mathrm{K}^+\mathrm{K}^- \to \pi^+\pi^-)}{(k-p-q)^2 - m_{\mathrm{K}}^2} + \tilde{F}_3 . \tag{55}$$

The transition amplitude $T(\mathrm{K}^+\mathrm{K}^- \to \pi^+\pi^-)$, partly off mass-shell, is estimated from current algebra and $(3, 3^*) + (3^*, 3)$ symmetry breaking [compare Eq. (50)] in a Weinberg-type expansion [14]:

$$T = \frac{\{m_K^2 + m_\pi^2 - (k^2 + q^2 + 2p^2 + 2pq - 2kp)\}}{2F^2}. \tag{56}$$

In arriving at Eq. (56), K-pole dominance had to be used. In this approximation, consistency of the various low-energy limits with a second-order expansion demands $F_\pi = F_K = F$. We will drop the distinction between F_π and F_K for the following discussion.

In the limit $p \to 0$, $q^2 = m_\pi^2$, the decay amplitude has to vanish, since the relevant commutator does not produce currents of isospin 3/2. We deduce

$$\tilde{F}_3 = 0 \quad \text{(coefficient of } k_\mu), \tag{57}$$

$$F_1 - F_2 - \tilde{F}_3 = 0 \quad \text{(coefficient of } q_\mu). \tag{58}$$

In the limit $q \to 0$, $p^2 = m_\pi^2$, the decay amplitude is related to the amplitude of K_{13} ecay. We deduce

$$\left(\tilde{F}_3 - \frac{1}{\sqrt{2}F}\right) = -\frac{1}{\sqrt{2}F}(F^+ + F^-) \quad \text{(coefficient of } k_\mu), \tag{59}$$

$$\left(F_1 + F_2 - \tilde{F}_3 + \frac{1}{\sqrt{2}F}\right) = -\frac{1}{\sqrt{2}F}(F^+ - F^-) \quad \text{(coefficient of } p_\mu). \tag{60}$$

Eq. (59) is equivalent to the Callan-Treiman relation (42); the other equations specify

$$F_1 \approx F_2 \approx -(F^+/\sqrt{2}F). \tag{61}$$

Further low-energy theorems can be deduced from the matrix element of the current divergence

$$\begin{aligned} &\langle \pi^+(p), \pi^-(q) | \partial A^{K^-} | K^+(k) \rangle \\ &\quad = (k-p-q)^\mu \{F_1(p+q)_\mu + F_2(p-q)_\mu + F_3(k-p-q)_\mu\}. \end{aligned} \tag{62}$$

In accordance with earlier approximations, it is related to the (πK)-scattering amplitude by K − PCAC

$$\langle \pi^+(p), \pi^-(q) | \partial A^{K^-} | K^+(k) \rangle = \frac{\sqrt{2}F \cdot m_K^2\, T(K^+K^- \to \pi^+\pi^-)}{(k-p-q)^2 - m_K^2}. \tag{63}$$

The construction of T in Eq. (55) automatically takes account of the possible low-energy theorems. The limit $p \to 0$ eliminates the possible addition of a constant, representing background to the pole term in (63). Substituting the current algebra estimates for the K_{14} form factors into

(62) and (63), we get the consistency requirement:

$$\frac{\sqrt{2}F \cdot m_K^2 \cdot T(K^+K^- \to \pi^+\pi^-)}{(k-p-q)^2 - m_K^2} = -\left((k-p-q)^\mu \cdot \frac{F^+}{\sqrt{2}F} 2p_\mu\right) + \frac{\sqrt{2}F(p+q-k)^2\, T(K^+K^- \to \pi^+\pi^-)}{(k-p-q)^2 - m_K^2}, \tag{64}$$

which reduces to

$$\sqrt{2}F\, T(K^+K^- \to \pi^+\pi^-) = -\frac{F^+}{\sqrt{2}F}(2m_\pi^2 + 2pq - 2kp) \tag{65}$$

with two pions and one K-meson on their mass shells. Comparison with (56) enforces $F^+ = 1$. It should be noted that the simultaneous requirement of π and K-pole dominance enforces SU_3-symmetric results: $F_\pi = F_K = F$, $F^+ = 1$ and as a consequence $F^- = 0$ [see (59)]. We will return to this point.

6. The Structure of Low Energy Theorems

We have discussed in detail the statements current algebra makes on K_{l3} and K_{l4} decay. Similar low-energy theorems can be derived for weak pion production and for nonleptonic decays [15].

We always meet the problem of extrapolating from an unphysical low-energy limit of a soft pion or K-meson amplitude to a physically accessible point. In some cases the assumption of smooth extrapolation is plausible (K_{l3} decay), in other cases certain strongly varying contributions are given a separate treatment (K_{l4} decay). A more systematic extrapolation, if practicable, is given by dispersion relations, where the corrections to the meson-pole terms appear explicitly in form of cuts [16].

A very successful method for extrapolating into the medium energy domain has been proposed by *Schnitzer* and *Weinberg* [17], the so-called hard meson calculations, which exploit simultaneously all the constraints placed by current algebra on generalized amplitudes involving *n* currents or less. Special solutions to the equations of hard meson analysis are provided by the tree graphs of chiral *Lagrangians* [18] which can be constructed to take account of the dynamical variations induced by any chosen set of particles in the pole approximation. Only sparse attempts have been made to go beyond pole diagrams.

It should be noted that the present dynamical framework of meson-pole dominance puts severe constraints on the structures which can be realized through current algebra. We have observed that a consistent simultaneous application of pion and K-meson pole dominance in

K_{14} leads to SU_3 symmetric results; it leaves no room for SU_3 violations. From the practical point of view one would regard this as an argument to abandon K − PCAC results in dealing with SU_3 violations. In this line of exploration, however, we would like to study the structure of meson-pole dominated dynamics.

Meson-pole dominance in itself is ambiguous, unless we add a dynamical postulate, which we will characterize below. We compare two methods of extrapolating two pions off mass shell.

$$\begin{aligned}(A)\ \langle\pi^i(q)\,a(p)|\pi^j(q')\,b(p')\rangle_{(A)} &= \frac{i(m_\pi^2-q^2)}{F_\pi m_\pi^2}\int e^{iqx}\langle a(p)|\partial A^i(x)|\pi^j(q')\,b(p')\rangle\,(\mathrm{d}x)^4\\ &= \frac{(m_\pi^2-q^2)}{F_\pi m_\pi^2}\,q^\mu\int e^{iqx}\langle a(p)|A^i_\mu(x)|\pi^j(q')\,b(p')\rangle\,(\mathrm{d}x)^4\\ &= \frac{i(m_\pi^2-q^2)\,(m_\pi^2-q'^2)}{(F_\pi m_\pi^2)^2}\,q^\mu\int e^{iqx}\,e^{-iq'y}\\ &\quad\cdot\langle a(p)|T\{A^i_\mu(x),\partial A^j(y)\}|b(p')\rangle\,(\mathrm{d}x)^4\,(\mathrm{d}y)^4\,;\end{aligned}\tag{66}$$

$$\begin{aligned}(B)\ \langle\pi^i(q)\,a(p)|\pi^j(q')\,b(p')\rangle_{(B)} &= -\frac{(m_\pi^2-q^2)\,(m_\pi^2-q'^2)}{(F_\pi m_\pi^2)^2}\int e^{iqx}\,e^{-iq'y}\\ &\quad\cdot\langle a(p)|T\{\partial A^i(x),\partial A^j(y)\}|b(p')\rangle\,(\mathrm{d}x)^4\,(\mathrm{d}y)^4\\ &= i\,\frac{(m_\pi^2-q^2)\,(m_\pi^2-q'^2)}{(F_\pi m_\pi^2)^2}\,q^\mu\int e^{iqx}\,e^{-iq'y}\\ &\quad\cdot\langle a(p)|T\{A^i_\mu(x),\partial A^j(y)\}|b(p')\rangle\,(\mathrm{d}x)^4\,(\mathrm{d}y)^4\\ &\quad+\frac{(m_\pi^2-q^2)\,(m_\pi^2-q'^2)}{(F_\pi m_\pi^2)^2}\int e^{i(q-q')y}\cdot\\ &\quad\cdot\langle a(p)|i\sigma^{ij}(y)|b(p')\rangle\,(\mathrm{d}y)^4\end{aligned}\tag{67}$$

with $[Q^i_A,\partial A^j(y)]=i\sigma^{ij}(y)$.

The two extrapolations (A) and (B) differ by the last term of Eq. (67), the so-called σ-term. The following alternatives present themselves:

(a) Pion-pole dominance is not satisfied in one or the other extrapolation.

(b) σ-terms are small and vanish in the limit of exact pion-pole dominance. In line with our original question we investigate this hypothesis.

We form:

$$\frac{\mathrm{d}}{\mathrm{d}t}[Q_A^i, Q_A^j] = [\int \partial A^i, Q_A^j] + [Q_A^i, \int \partial A^j] = i f^{ijk} \frac{\mathrm{d}}{\mathrm{d}t} Q_V^k = i f^{ijk} \int \partial V^k \tag{68}$$

and omitting the space integration, we have:

$$i(\sigma^{ij} - \sigma^{ji}) = i f^{ijk} \partial V^k . \tag{69}$$

We propose the following interpretation: Since at least to our present knowledge we have no grounds for prefering the successive or the simultaneous extrapolation of the two pions, we have to accept the effects of σ-terms as providing an inherent ambiguity in our use of meson-pole dominance. Eq. (69) then shows that this ambiguity covers any effects associated with possible nonconservation of the vector current in the commutator. This is to be spelled out in two cases of interest:

(I) Let us return to the stage where the commutator of two axial charges has not yet been specified. On general grounds one can only conclude that it is some vector charge of isospin one. To the extent that pion-pole dominance is universally valid, the σ-terms are neglected and the vector charge is conserved. By lack of other conserved charges, it is set proportional to the isospin charge

$$[Q_A^i, Q_A^j] = i\, \mathrm{e}^{ijk} \alpha I . \tag{70}$$

If there are corrections to (70), we would not be able to separate them from PCAC extrapolation errors.

The scale of the commutator cannot be specified by arguments of this type; it is prescribed by weak interaction universality to be $\alpha = 1$.

(II) In a similar way we can now argue that the simultaneous use of π and K-meson pole dominance obscures any features associated with the non-conservation of the strangeness changing vector current in their current-commutator. This is how the emergence of SU_3 symmetry results in our analysis of K_{l4} decay should be understood, and this is the deeper reason why K-pole dominance should not be used in analysing SU_3 violations.

From the required smallness of σ-terms we can make further deductions. We apply a Jacobi identity [with $\alpha = 1$ in (70)]:

$$\begin{aligned} i[Q_A^i, \sigma^{jk}] &= [Q_A^i, [Q_A^j, \partial A^k]] \\ &= -[\partial A^k, [Q_A^i, Q_A^j]] - [Q_A^j, [\partial A^k, Q_A^i]] \\ &= \delta_{jk} \partial A^i - \delta_{ik} \partial A^j + i[Q_A^j, \sigma^{ik}] . \end{aligned} \tag{71}$$

In Eq. (71) we see that in the limit where σ-terms are disregarded, the axial-vector divergence would vanish. Conservation of the axial charge implies that eigenstates of the energy operator are parity degenerate, this is achieved either by having parity doublets or massless pseudoscalar bosons. In our case, this limit is realized by taking $m_\pi \to 0$ with $F_\pi \neq 0$ so that pion-pole dominance is maintained. In a recent paper, *Dashen* and *Weinstein* [11] have given detailed examples to demonstrate how the usual soft-pion theorems become exact in that limit.

From Eq. (71) we recognize: to the extent that pion-pole dominance is imposed and corrections to this approximation are disregarded, the low-energy theorems of current algebra reproduce structure of chiral symmetry. (For more extensive arguments compare Ref. [11].)

7. The Open Questions

We have seen that the current algebra results fit in very smoothly with strong interaction dynamics in the low-energy domain, which is governed by the dominance of simple meson poles. To the extent that meson-pole dominance is valid, the currents are constrained to exhibit a simple symmetry algebra, and not much scope is available for alternative algebraic schemes.

These constraints can be described either as an effect of strong interaction dynamics on the possible structures of a current algebra, as we did, or as an implication of current algebra on the structure of strong interactions [18]. In a theory where Eq. (70) is not approximately valid (like the Yukawa model) or where no chiral symmetry limit is at hand, the approximation of meson pole dominance should be discouraged. This would be the necessary consequence if the weak and electromagnetic currents did not turn out to have the simple algebraic structure under discussion.

Continuing this type of consideration, we may eventually be able to justify certain aspects of current algebra from hadron dynamics; but we will not be able to deduce the scales which are postulated through current algebra for the weak currents. Note that the scale of a current in a hadron symmetry algebra is not necessarily identical to its scale in a weak coupling to leptons. As long as we describe weak interactions only as first-order transitions, this description will not provide a deeper justification of the scale of the coupling either. So it is necessary to go to higher-order weak interactions for a deeper understanding. Whether this will provide the desired insight, remains to be seen.

Acknowledgement. I wish to thank Prof. J. Wess for his hospitality at Karlsruhe during the summer school, Prof. H. Lehmann for his hospitality at Hamburg, where these lectures were prepared, and Dr. N. J. Carron for his valuable help in writing these lecture notes.

References

1. *Gell-Mann, M.:* Phys. Rev. **125**, 1067 (1962).
2. See the lectures by *H. Primakoff* (this school).
3. See the lectures by *V. Müller* (this school).
4. *Adler, S. L.:* Phys. Rev. **135**, B 963 (1964).
5. — Phys. Rev. **143**, 1144 (1966); E **151**, 1342 (1966).
6. —, *Gilman, F. J.:* Phys. Rev. **156**, 1598 (1967).
7. See the lecture by *J. Steinberger* at the Lund International Conference on Elementary Particles (1969).
8. However there are substantial difficulties in achieving a quantitative explanation of this correction; see *Pagels, H.:* Phys. Rev. **179**, 1337 (1969).
9. *Adler, S. L.:* Phys. Rev. Letters **14**, 1051 (1965) and Phys. Rev. **140**, B 736 (1965); E **149**, 1294 (1966), E **175**, 2224 (1968) – *Weisberger, W. I.*, Phys. Rev. Letters **14**, 1047 (1965) and Phys. Rev. **143**, 1302 (1966).
10. This can be shown on very general grounds, see for instance: *Haag, R.:* Phys. Rev. **112**, 669 (1958).
11. See for instance: *Weinberg, S.:* 14th International Conference on High Energy Physics, Vienna 1968, Proceedings p. 253 or *Dashen, R., Weinstein, M.:* Phys. Rev. **183**, 1261 (1969).
12. *Callan, C. G., Treiman, S. B.:* Phys. Rev. Letters **16**, 153 (1966); *Mathur, V. S., Okubo, S., Pandit, L. K.:* Phys. Rev. Letters **16**, 371 (1966).
13. *Gell-Mann, M., Oakes, R. J., Renner, B.:* Phys. Rev. **175**, 2195 (1968). *Ellis, J.:* Nucl. Phys. B **13**, 153 (1969).
14. *Griffith, R. W.:* Phys. Rev. **176**, 1705 (1968); see also: — *Renner, B.:* Physics Letters **27** B, 650 (1968).
15. See the lectures by *B. Stech* (this school).
16. See for instance *Fubini, S., Furlan, G.:* Ann. Phys. (NY) **48**, 322 (1968).
17. *Schnitzer, H., Weinberg, S.:* Phys. Rev. **164**, 1828 (1967).
18. For a review with references, see *Weinberg, S.*, Ref. [11].

Dr. *B. Renner*
Department of Applied Mathematics
and Theoretical Physics
University of Cambridge
Cambridge/England

The Decay of the K_0—$\bar{K}_0$ System

H. G. Dosch

Contents

I. Introduction 79
II. The Weisskopf-Wigner Method 80
III. The $K_0 - \bar{K}_0$ System 85
IV. The Exponential Decay 87
References and Footnotes 90

I. Introduction

In the following we will discuss the problem of decaying quantum-mechanical states with particular emphasis on the decay of the neutral K-mesons.

These pseudoscalar mesons decay by weak interactions into the following states: two pions, three pions and one pion plus a lepton pair. Since the quantum number which allows us to distinguish the K_0 from its antiparticle $\bar{K}_0$, namely the strangeness, is not conserved in weak interactions, this quantum number is in general no longer useful for the description of the decaying states. One therefore encounters the well known mixing of particles and antiparticles which leads to the long and short lived component of the neutral K-mesons.

We first formulate the problem which will be treated in the framework of nonrelativistic quantum mechanics.

There is a Hamiltonian H which describes the time dependence of the state $\psi(t)$

$$i\,\partial_t\,\psi(t) = H\,\psi(t)\,. \tag{1}$$

This Hamiltonian shall consist of two parts

$$H = H_0 + H_1 \tag{2}$$

a "free" part H_0 and a "small" interaction part H_1. If we speak of a K meson, a 3π state etc. we mean eigenstates of the Hamiltonian H_0.

These eigenstates form a complete set and we can expand any state $\psi(t)$ in terms of the states ψ_n:

$$\psi(t) = \sum_{n=1}^{\infty} a_n(t)\,\psi_n = \sum_{n=1}^{\infty} b_n(t)\,\mathrm{e}^{-iE_n t}\,\psi_n\,, \tag{3}$$

$$H_0\,\psi_n = E_n\,\psi_n\,. \tag{3a}$$ *

From Eqs. (3) and (1) follow the differential equations

$$i\,\partial_t b_n(t) = \sum_{n'} H_{nn'} b_{n'}(t)\,\mathrm{e}^{i(E_n - E_{n'})t}\,, \qquad n = 1, 2, \ldots \tag{4}$$

with

$$H_{nn'} = (\psi_n, H_1\,\psi_{n'})\,. \tag{4a}$$

We now identify the first two states ψ_1 and ψ_2 with the particles, the decay of which we are interested in. We are only interested in the functions $b_1(t)$ and $b_2(t)$ which describe the time dependence of the decaying states, and we are not interested in what happens to the decay products. Therefore one wants to find a new set of equations yielding just this information, but simpler than Eq. (4). This has been done first by *Weisskopf* and *Wigner* in 1931 [1]. The simplest way to proceed is to assume, that the time derivatives of the functions $b_1(t)$ and $b_2(t)$ are linear functions of the amplitudes themselves:

$$i\,\partial_t b_i(t) = \sum_{k=1}^{2} R_{ik}\,b_k(t) \tag{5}$$

where the elements of the general 2×2 matrix R have to be determined by experiment. One normally goes to the frame where the decaying particles are at rest, the time t is then the eigentime.

One can also be more ambitious and try to derive Eq. (5). A heuristic and rather instructive way, which allows us to express the matrix elements R_{ik} in terms of the matrix elements of the interaction Hamiltonian H_1 starts from a perturbation theoretical solution of Eq. (4). We will discuss this way in some detail in the next section. In Section III we will apply the method to the $K_0\bar{K}_0$ system and give in Section IV a more direct approach to the problem of decaying states.

II. The Weisskopf-Wigner Method

The integrated form of Eqs. (4), to wit

$$b_n(\Delta t) - b_n(0) = \frac{1}{i}\int_0^{\Delta t} \mathrm{d}t' \sum_{n'=1}^{\infty} H_{nn'}\,\mathrm{e}^{i(E_n - E_{n'})t'}\,b_{n'}(t') \tag{6}$$

---

* This is valid for a finite volume. If one chooses an infinite normalization volume and ψ_n is a continuum state, Eq. (3a) is to be understood in the sense that ψ_n is a eigenpacket with its energy sharply peaked around E_n.

can be used in order to obtain an iterative series:

$$b_n^{(i+1)}(\Delta t) - b_n(0) = \frac{1}{i}\int_0^{\Delta t} dt' \sum_{n'=1}^{\infty} H_{nn'}\, e^{i(E_n - E_{n'})t'}\, b_{n'}^{i}(t'), \tag{6a}$$

$$b_n^{(0)}(t) = b_n(0),$$

where the superscript gives the order of iteration. We now confine ourselves to the second order, choosing the initial conditions

$$b_1(0) \neq 0; \quad b_2(0) \neq 0; \quad b_n(0) = 0 \quad n = 3, \ldots$$

Replacing the summation over the continuum states ($n \geqq 3$) by an integration over the energy we obtain:

$$b_i^{(2)}(\Delta t) - b_i(0) = I_1 + I_2 + I_3 \tag{7}$$

with:

$$I_1 = -\sum_{k=1}^{2} \frac{e^{i(E_i - E_k)\Delta t} - 1}{E_i - E_k}\, H_{ik} b_k(0) \tag{7a}$$

$$I_2 = \sum_{k,j=1}^{2} H_{ij} H_{jk} \left\{ \frac{e^{i(E_i - E_k)\Delta t} - 1}{(E_j - E_k)(E_i - E_k)} - \frac{e^{i(E_i - E_j)\Delta t} - 1}{(E_j - E_k)(E_i - E_j)} \right\} b_k(0), \tag{7b}$$

$$I_3 = \sum_{k=1}^{2} \int_{E_0}^{\infty} dE\, X_{ik}(E) \left\{ \frac{e^{i(E_i - E_k)\Delta t} - 1}{(E - E_k)(E_i - E_k)} - \frac{e^{i(E_i - E)\Delta t} - 1}{(E - E_k)(E_i - E)} \right\} b_k(0) \tag{7c}$$

where

$$X_{ik}(E) = \sum_{\alpha} \varrho^{\alpha}(E)\, H_{i,E_\alpha} H_{E_\alpha,k}, \tag{8}$$

$$H_{i,E_\alpha} = (\psi_i, H_1 \psi_{E_\alpha}). \tag{8a}$$

The index α specifies different states of the same energy, it distinguishes for instance a 2π from a 3π state.

In order to get a differential equation like Eq. (5) we have to obtain linearity of $b_i(\Delta t) - b_i(0)$ in Δt for small times Δt. We therefore discuss the integrals I_1, I_2, and I_3 under a number of assumptions which lead to linearity in Δt.

We first choose Δt such that:

$$(E_i - E_k)\,\Delta t \ll 1. \tag{a1}$$

Then higher terms in a power series expansion of $\exp\{i(E-E)\Delta t\}$ can be neglected and one obtains:

$$I_1 = -i\Delta t \sum_{k=1}^{2} H_{ik} b_k(0), \tag{9}$$

$$I_2 = -\frac{1}{2}(\Delta t)^2 \sum_{j,k=1}^{2} H_{ij} H_{jk} b_k(0). \tag{10}$$

We assume further that

$$\Delta t \cdot H_{ij} \ll 1 \tag{a2}$$

and hence the non-linear term I_2 can be neglected in comparison with I_1.

The third term I_3 is split into three parts:

$$I_3 = I_{31} + I_{32} + I_{33}, \tag{11}$$

$$I_{31} = \int_{E_0}^{E_1-\eta} \mathrm{d}E(\mathrm{Int}); \qquad I_{32} = \int_{E_1+\eta}^{\infty} \mathrm{d}E(\mathrm{Int}), \tag{11a}$$

$$I_{33} = \int_{E_1-\eta}^{E_1+\eta} \mathrm{d}E(\mathrm{Int}), \tag{11b}$$

$$\mathrm{Int} = \sum_{k=1}^{2} X_{ik}(E) \left\{ \frac{\mathrm{e}^{i(E_i-E_k)\Delta t}-1}{(E-E_k)(E_i-E_k)} - \frac{\mathrm{e}^{i(E_i-E)\Delta t}-1}{(E-E_k)(E_i-E)} \right\} b_k(0). \tag{11c}$$

We now make the following assumptions:

There is an interval η such that the variation of $X_{ik}(E)$ in the interval $(E_1-\eta, E_1+\eta)$ can be neglected, with (a 3)

$$\eta\,\Delta t \gg 1, \tag{a 4}$$

$$|E_1 - E_2| \ll \eta. \tag{a 5}$$

Because of the condition (a 4) one can neglect the second term in the integrand of I_{31} and I_{32} since it is smaller than the first term by at least the factor $1/(\Delta t \cdot \eta)$. Therefore we obtain:

$$\begin{aligned} I_{31} + I_{32} &\approx (i\Delta t)\left(\int_{E_0}^{E_1-\eta} + \int_{E_1+\eta}^{\infty}\right) \mathrm{d}E \sum_{k=1}^{2} \frac{X_{ik}(E)}{E-E_k} b_k(0) \\ &\approx (i\Delta t)\,\mathrm{P}\int_{E_0}^{\infty} \mathrm{d}E \sum_{k=1}^{2} \frac{X_{ik}(E)}{E-E_k} b_k(0). \end{aligned} \tag{12}$$

The last approximate equality is a consequence of assumption (a 3).

Because of condition (a 3) the function $X_{ik}(E)$ can be taken out of the integral and after some rearrangements in the curly brackets one obtains:

$$I_{33} = \sum_{k=1}^{2} X_{ik}(E_1)\, b_k \int_{(E_1-\eta)\Delta t}^{(E_1+\eta)\Delta t} \mathrm{d}(E\,\Delta t) \frac{1}{E_i - E_k} \times \left\{ \mathrm{e}^{i(E_i-E_k)\Delta t} \frac{\mathrm{e}^{i(E_k-E)\Delta t}-1}{(E_k-E)\,\Delta t} - \frac{\mathrm{e}^{i(E_i-E)\Delta t}-1}{(E_i-E)\,\Delta t} \right\}. \tag{13}$$

Using assumption (a 4) and the Dirichlet Integral

$$\int_{-\infty}^{+\infty} \mathrm{d}y \frac{\mathrm{e}^{iy}-1}{y} = i\pi$$

one finds the simple expression:

$$I_{33} \approx -\pi\,\Delta t \sum_{k=1}^{2} X_{ik}(E_1)\, b_k(0)\,. \tag{13a}$$

Now Eqs. (12) and (13a) can be combined to

$$I_3 = i\,\Delta t \int_{E_0}^{\infty} \mathrm{d}E \frac{X_{ik}(E)}{E - E_k - i\varepsilon}\, b_k(0)\,. \tag{14}$$

One should note that the relation

$$\frac{1}{x + i\varepsilon} = P\frac{1}{x} - i\pi\,\delta(x)$$

has only been used in order to write the result in a concise form and not in order to derive Eq. (12) and (13). For the derivation of these equations we had to rely heavily on assumptions (a 3) through (a 5).

We now collect our results and obtain:

$$i\frac{b_i^{(2)}(\Delta t) - b_i(0)}{\Delta t} = \sum_{k=1}^{2} R_{ik} b_k(0) \tag{15}$$

with

$$R_{ik} = H_{ik} - \int_{E_0}^{\infty} \mathrm{d}E \frac{X_{ik}(E)}{E - E_k - i\varepsilon}\,. \tag{16}$$

This linear approximation is certainly only valid if the initial condition is not too much perturbed, i.e. if

$$|b_i^{(2)}(\Delta t) - b_i(0)| \ll b^i(0). \tag{a6}$$

If we had started at a later time t_0 and had computed the difference $b_i^{(2)}(\Delta t + t_0) - b_i(t_0)$, the main difference would have been, that the initial conditions would be different. One could no longer assume, that $b_n(t_0) = 0$ for $n \geqq 3$. In order to deal with this problem we best take resource to our physical intuition and convince ourselves, that the recombination of decay products to the decaying particle is negligible [2].

Therefore one would obtain a similar relation:

$$i\,\frac{b_i^{(2)}(t_0 + \Delta t) - b_i(t_0)}{\Delta t} = \sum_{k=1}^{2} R_{ik} b_k(t_0). \tag{15a}$$

In the limit $\Delta t \to 0$ one obtains from Eq. (15) or (15a) the desired differential equation

$$i\,\partial_t b_i(t) = \sum_{k=1}^{2} R_{ik} b_k(t), \tag{5}$$

with the non hermitian matrix R given by Eq. (16).

Since condition (a 3) gives a lower limit for Δt, the solutions of the differential Eq. (5) will only approximate the solutions of the exact problem, if the latter ones are averaged over time intervals Δt such that $\eta\,\Delta t \gg 1$.

It should be noted, that conditions (a 1), (a 2) and (a 6) which gave upper limits for Δt were used in order to derive the difference Eq. (15). The solutions of the differential equation, however, are supposed to be valid for much longer time intervals (see Section IV).

Another way to proceed [3] is to insert the solutions of Eq. (5) into the Schrödinger equation (4). Under the assumption that the matrix elements $H_{nn'}$ vanish if not at least one index has the value 1 or 2, one verifies that the solutions of Eq. (5) are also approximate solutions of Eq. (4) and one also obtains for the matrix R the expression given in Eq. (16).

If $(b_1^{\mathrm{I}}(0), b_2^{\mathrm{I}}(0)) = \boldsymbol{b}^{\mathrm{I}}$ and $(b_1^{\mathrm{II}}(0), b_2^{\mathrm{II}}(0)) = \boldsymbol{b}^{\mathrm{II}}$ are eigenvectors of R, i.e.

$$\sum_{k=1}^{2} R_{ik} b_k^N = r_N b_i^N \qquad r_N = -(\delta_N + i\gamma_N), \qquad N = \mathrm{I}, \mathrm{II} \tag{17}$$

(r_N is complex since R is not hermitian), the time dependence is particularly simple:

$$\boldsymbol{b}^N(t) = \boldsymbol{b}^N(0)\, e^{-i r_N t}, \qquad N = \mathrm{I}, \mathrm{II}. \tag{18}$$

The eigenvalues r_N in terms of the elements of R are:

$$r_{\mathrm{I,II}}=\frac{R_{11}+R_{22}}{2}\pm D\,;\quad D=\tfrac{1}{2}\{(R_{11}-R_{22})^2+4R_{12}R_{21}\}^{1/2}\,. \quad (19)$$

If the eigenvectors $\boldsymbol{b}^N$ are written in the form:

$$\boldsymbol{b}^{\mathrm{I}}=\frac{1}{\sqrt{2}(1+|\varepsilon_{\mathrm{I}}|^2)^{1/2}}\begin{pmatrix}1+\varepsilon_{\mathrm{I}}\\ 1-\varepsilon_{\mathrm{I}}\end{pmatrix};\quad \boldsymbol{b}^{\mathrm{II}}=\frac{1}{\sqrt{2}(1+|\varepsilon_{\mathrm{II}}|^2)^{1/2}}\begin{pmatrix}1+\varepsilon_{\mathrm{II}}\\ -1+\varepsilon_{\mathrm{II}}\end{pmatrix}, \quad (20)$$

the quantities ε_{I} and $\varepsilon_{\mathrm{II}}$ are:

$$\varepsilon_{\mathrm{I}}=\frac{R_{12}-R_{21}+R_{11}-R_{22}}{R_{12}+R_{21}+2D}, \quad (21\mathrm{a})$$

$$\varepsilon_{\mathrm{II}}=\frac{R_{12}-R_{21}-R_{11}+R_{22}}{R_{12}+R_{21}+2D}. \quad (21\mathrm{b})$$

Sometimes the matrix $m_i\delta_{ik}+R_{ik}$ is split into a hermitian "mass matrix" M_{ik} and an anti-hermitian "decay-matrix" $i\Gamma_{ik}$:

$$m_i\delta_{ik}+R_{ik}=M_{ik}+i\Gamma_{ik}\,. \quad (22)$$

III. The K_0—$\bar{K}_0$ System

We identify the state ψ_1 with the K_0 and ψ_2 with the $\bar{K}_0$ state and go to their rest system. Inserting the definition (8) into Eq. (16) one obtains for the elements of R:

$$R_{11}=(\psi_{\mathrm{K}},H_1\psi_{\mathrm{K}})-\int_{E_0}^{\infty}\mathrm{d}E\,\frac{\varrho(E)\,|(\psi_E,H_1\psi_{\mathrm{K}})|^2}{E-m_{\mathrm{K}}-i\varepsilon}, \quad (23\mathrm{a})$$

$$R_{22}=(\psi_{\bar{\mathrm{K}}},H_1\psi_{\bar{\mathrm{K}}})-\int_{E_0}^{\infty}\mathrm{d}E\,\frac{\varrho(E)\,|(\psi_E,H_1\psi_{\bar{\mathrm{K}}})|^2}{E-m_{\bar{\mathrm{K}}}-i\varepsilon}, \quad (23\mathrm{b})$$

$$R_{12}=(\psi_{\mathrm{K}},H_1\psi_{\bar{\mathrm{K}}})-\int_{E_0}^{\infty}\mathrm{d}E\,\frac{\varrho(E)\,(\psi_{\mathrm{K}},H_1\psi_E)\,(\psi_E,H_1\psi_{\bar{\mathrm{K}}})}{E-m_{\bar{\mathrm{K}}}-i\varepsilon}, \quad (23\mathrm{c})$$

$$R_{21}=(\psi_{\bar{\mathrm{K}}},H_1\psi_{\mathrm{K}})-\int_{E_0}^{\infty}\mathrm{d}E\,\frac{\varrho(E)\,(\psi_{\bar{\mathrm{K}}},H_1\psi_E)\,(\psi_E,H_1\psi_{\mathrm{K}})}{E-m_{\mathrm{K}}-i\varepsilon}. \quad (23\mathrm{d})$$

We first review the assumptions (a 1) through (a 6):

(a 1) The mass difference between the K_0 and the $\bar{K}_0$ is presumably zero but certainly smaller than 10^{-12} MeV and therefore one deduces from (a 1): $\Delta t \ll 10^{-10}$ sec.

(a 2) We can add the diagonal terms $(\psi_i, H_1\psi_i)$ to the free Hamiltonian H_0 and are left only with the condition $\Delta t |(\psi_K, H_1\psi_{\bar{K}})| \ll 1$ which again leads to $\Delta t \ll 10^{-10}$ sec.

(a 3) The assumption $\eta \approx 1$ MeV is certainly very conservative.

(a 4) Follows from (a 3) to demand $t \gg 10^{-22}$ sec.

(a 5) Is also satisfied (see (a 1) and (a 3)).

(a 6) Again leads to $t \ll 10^{-10}$ sec.

So we see that the above conditions can be fulfilled in a selfconsistent way.

Finally we want to investigate the consequences of discrete symmetries on the matrix R. We consider the following operations: The charge conjugation C, which transforms a particle into its anti-particle, the parity operation P and the time reversal T. We have good reasons to believe that the product CPT commutes [4] with H from which follows:

$$R_{11} = R_{22}. \tag{24}$$

This can be seen in the following way:

$$\begin{aligned}(\psi_E, H_1\psi_K) &= (CPT\psi_E, CPTH_1\psi_K)^* \\ &= (CPT\psi_E, H_1 CPT\psi_K)^* = (CPT\psi_E, H_1\psi_{\bar{K}})^*\end{aligned} \tag{24a}$$

(the complex conjugation is due to the anti-unitarity of T).

Inserting the identity (25) in Eq. (22a) one remarks, that the complex conjugation is of no influence and furthermore, since the summation goes over a complete set, the change from the basis ψ_E to $CPT\psi_E$ does not matter. One also sees immediately that if H commutes with CPT, one has

$$m_K = m_{\bar{K}} \quad \text{and} \quad (\psi_K, H_1\psi_K) = (\psi_{\bar{K}}, H_1\psi_{\bar{K}}).$$

Then the identity (24) follows and from comparison with Eqs. (21) one obtains:

$$\varepsilon_{\mathrm{I}} = \varepsilon_{\mathrm{II}}. \tag{25}$$

If furthermore the operation CP would commute with H one would obtain $R_{12} = R_{21}$ and hence $\varepsilon_{\mathrm{I}} = \varepsilon_{\mathrm{II}} = 0$.

IV. The Exponential Decay

In this section the problem of a decaying state is investigated from a different point of view. We confine ourselves to the treatment of only one decaying state. This simplifies the technique considerably.

The time dependence of the coefficient $b_1(t)$ is given by:

$$b_1(t) = e^{iE_1 t}(\psi_1, e^{-iHt}\psi_1). \tag{26}$$

We now follow the method of *Höhler* [5], by using an integral representation for $\exp(-iHt)$ and splitting up the resulting expressions for Eq. (26) into a term which falls off exponentially and a term which gives the deviations from the exponential decay.

The integral representation used in Ref. [5] is:

$$F(H) = \frac{1}{2\pi i} \int_C d\lambda \frac{F(\lambda)}{\lambda \cdot 1 - H} d\lambda. \tag{27}$$

The contour C encloses the whole spectrum of H, i.e. the resolvent $(\lambda \cdot 1 - H)^{-1}$ exists by definition on and outside of C. The representation (27) is valid if $F(\lambda)$ is regular inside the contour C.

Since the spectrum of H is not bound and since the exponential function has an essential singularity at infinity the representation (27) can not be used directly. One way to overcome this difficulty is to truncate the spectrum of H, i.e. to introduce the self-adjoint operator H_N

$$H_N = \int_{E_0}^{N} \lambda \, dE_\lambda \tag{28}$$

and to postulate that $\exp(-iH_N t)$ converges weakly to $\exp(-iHt)$. For H_N the representation (27) can be used and so we assume in the following always that H is a bounded operator with the upper bound N sufficiently large.

If one makes the assumption [6] that

$$(\psi_i, H_1 \psi_k) = 0 \quad \text{except for } i \text{ or } k = 1, \tag{29}$$

one can calculate easily the relevant matrix element

$$(\psi_1, (\lambda 1 - H)^{-1} \psi_1)$$

using the identity

$$(\lambda \cdot 1 - H_0)(\lambda 1 - H)^{-1} - H_1(\lambda \cdot 1 - H)^{-1} = 1.$$

For the function $b_1(t)$ one obtains:

$$b_1(t) = \frac{1}{2\pi i} \int_C \frac{e^{i(E_1 - \lambda)t}}{\lambda - E_1 + \Gamma(\lambda)} d\lambda \tag{30}$$

with

$$\Gamma(\lambda) = \sum_{n=2}^{\infty} \frac{|H_{1n}|^2}{E_n - \lambda}. \tag{31}$$

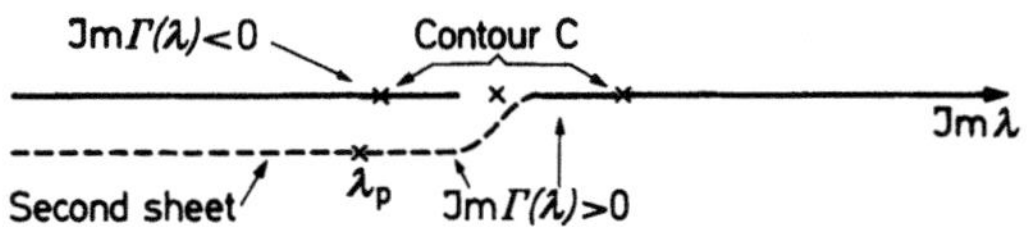

Fig. 1. Position of λ_p, the pole of the integrand of Eq. (30)

If again the summation over eigenpackets is replaced by an integration over the energy Eq. (31) becomes:

$$\Gamma(\lambda) = \int_{E_0}^{N} dE \frac{X(E)}{E - \lambda}; \quad X(E) = \sum_{\alpha} \varrho^{\alpha}(E) |H_{1,E_\alpha}|^2 \geqq 0. \tag{32}$$

If Γ were equal to zero (i.e. $H_1 = 0$) one would only get a contribution from the pole at $\lambda = E_1$ and would obtain:

$$b_1(t) = 1.$$

Since $\Gamma(\lambda)$ is assumed to be small but different from zero the pole will move away from $\lambda = E_1$. We now discuss the migration of this pole in some detail. The denominator of the integrand (30) on the upper respectively lower side of the contour ($\lambda = \lambda_0 \pm i\varepsilon$) is

$$\text{Den} = \lambda - E_1 + \mathrm{P} \int_{E_0}^{N} dE \frac{X(E)}{E - \lambda_0} \pm i\pi X(\lambda_0). \tag{33}$$

Therefore, the denominator has a positive imaginary part for $\operatorname{Im}\lambda > 0$ and a negative one if $\operatorname{Im}\lambda < 0$. Hence it cannot have a zero near the real axis if $\Gamma(\lambda)$ is given by the integral representation (32). This integral representation defines a function analytic in the cut λ plane. If $\Gamma(\lambda)$ is continued through the cut to values with $\operatorname{Im}\lambda < 0$ and a not too rapid variation of $\Gamma(\lambda)$ is assumed, this continued function, $\Gamma^{\mathrm{II}}(\lambda)$, has a positive imaginary part for $\operatorname{Im}\lambda < 0$ and therefore a zero of the denominator on the second sheet of the integrand is possible. The position of the zero of the denominator on the second sheet will be given by

$$\lambda_p = E_1 - \Gamma^{\mathrm{II}}(\lambda). \tag{34}$$

If $\Gamma(\lambda)$ is small, λ_p will be approximately E_1 and if $\Gamma^{II}(\lambda)$ varies not too fast near $\lambda = E_1$ we have approximately

$$\lambda_p \approx E_1 - \Gamma^{II}(E_1) = E_1 - \mathrm{P}\int_{E_0}^{\infty} \mathrm{d}E \frac{X(E)}{E - E_1} - i\pi X(E_1). \tag{34a}$$

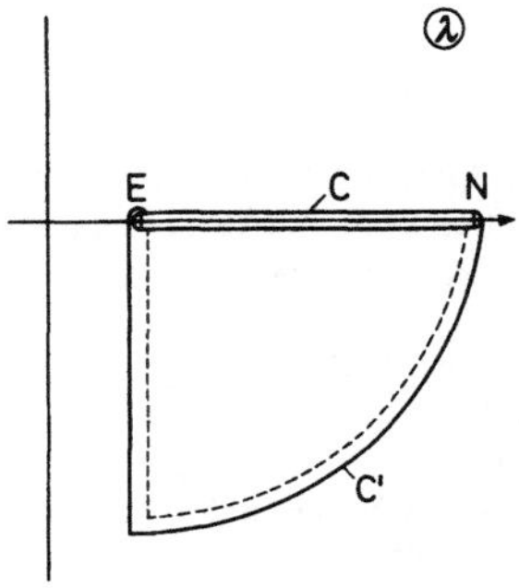

Fig. 2. Original and deformed contour of the integral (30)

Now the integration contour C is deformed to C' in such a way that the new contour C' encircles the pole on the second sheet, as indicated in Fig. 2 (the dashed integration path is on the second sheet). One obtains:

$$b_1(t) = \frac{\mathrm{e}^{i(E_1 - \lambda_p)t}}{1 + \frac{\mathrm{d}}{\mathrm{d}\lambda}\Gamma^{II}(\lambda_p)} + \frac{1}{2\pi i}\int_Q \frac{\{\Gamma^{II}(\lambda) - \Gamma(\lambda)\}\,\mathrm{e}^{i(E_1 - \lambda)t}}{\{\lambda - E_1 + \Gamma(\lambda)\}\,\{\lambda - E_1 + \Gamma^{II}(\lambda)\}}\,\mathrm{d}\lambda \tag{35}$$

where the integration *path* Q goes from E_1 to $E_1 + iN$ and from $E_1 + iN$ in a quartercircle to N.

The contribution of the pole has the familiar time dependence

$$\mathrm{e}^{i(E_1 - \lambda_q)t} = \mathrm{e}^{i\delta t - \gamma t}$$

with (36)

$$\delta = \mathrm{P}\int_{E_0}^{\infty} \mathrm{d}E \frac{X(E)}{E - E_1}; \qquad \gamma = \pi X(E_1) \qquad \text{(cf. Eq. (16))}.$$

The contribution from the cut can only be calculated if one has detailed information about $X(E)$ in order to perform the analytic continuation. It can be seen directly, however, that if H is proportional to a constant G (in our case the weak coupling constant) the cut contribution will be proportional to G^2.

Höhler [5] has estimated the cut contribution in the Lee model and has found for weak coupling that even though it does not fall off exponentially but rather like $t^{-\frac{3}{2}}$ it is so small that it only competes with the exponential term if the state has decayed practically completely (after more than 10 life times).

References and Footnotes

1. *Weisskopf, V., Wigner, E.:* Z. Physik **63**, 54 (1930); **65**, 18 (1930).
2. In order to clarify this point one had to make a calculation where the decay products are properly treated as wave packets and where the finite range of the interaction is taken into account.
3. See e.g. *Bernstein, J.:* In: *Lévy, M.* (Ed.): Cargese lectures in physics, Vol. 1. Gordon and Breach: New York 1966.
4. See e.g. *Streater, B. F., Wightman, A. S.: PCT,* spin and statistics and all that. New York: Dover Publ. 1964; Dt. Übers. B-I-Hochschultaschenbücher, Mannheim: 1969.
5. *Höhler, G.:* Z. Physik **152**, 546 (1958).
6. This assumption can be avoided, see *N. Byers, S. W. MacDowell,* and *C. N. Yang* in High Energy Physics and Elementary Particles; International Atomic Energy Agency (Triest 1965).

Dr. *H. G. Dosch*
Kernforschungszentrum Karlsruhe and
Institut für Hochenergiephysik der Universität
D-6900 Heidelberg, Albert-Überle-Straße 2

Questions Raised by CP-Nonconservation

P. K. KABIR

Contents

The Phenomenological Description 91
A Simple Model 96
$K^0 \to 2\pi$ Decays 97
Tests of T-Invariance 100
The Dynamical Origin of CP-Noninvariance 102
Superweak Theories 103
Milliweak Theories 104
Millistrong Theories 105
a-particles and Related Chimera 107
Summary 109
Appendix A 110
References and Footnotes 112

These lectures are divided roughly equally between phenomenological analysis of neutral kaon decays, and speculations concerning the unknown origin of CP-noninvariance. In the first part, in view of the uncertain experimental situation, we concentrate on very general considerations, which should be unaffected by changes in the experimental data. Should even those hypotheses prove to be invalid, that would be a discovery of at least equal importance. In the second, we classify the main possibilities, and then choose a few for more detailed description.

The Phenomenological Description

Because of the degeneracy or near-degeneracy of K^0 and $\bar{K}^0$ states, a state prepared initially as K^0 or $\bar{K}^0$ is not expected to decay in the simple exponential manner characteristic of unstable particles. Instead, certain linear superpositions of K^0 and $\bar{K}^0$ states may possess this property; the theoretical basis for the assumption of exponential decay was discussed in Dr. *Dosch*'s lectures.

We therefore simply assume that there exist two distinct superpositions of K^0 and $\bar{K}^0$ states [1]

$$|S\rangle = p_1 |K^0\rangle + q_1 |\bar{K}^0\rangle\,, \tag{1a}$$

$$|L\rangle = p_2 |K^0\rangle - q_2 |\bar{K}^0\rangle\,, \tag{1b}$$

which evolve exponentially with time, i.e.

$$|S\rangle \to \theta_1 |S\rangle\,, \qquad L \to \theta_2 |L\rangle \tag{2a}$$

with

$$\theta_j = \mathrm{e}^{-i\lambda_j t}\,, \qquad \lambda_j = m_j - \tfrac{1}{2} i\gamma_j\,. \tag{2b}$$

The particular superpositions $|S\rangle$ and $|L\rangle$ therefore each have the property that the *relative* amplitudes of $\bar{K}^0$ and K^0 components does not change with time. $|K^0\rangle$ and $|\bar{K}^0\rangle$ are taken to be orthogonal normalised states which would be eigenstates of the strong Hamiltonian if decay interactions did not exist. If we normalise S and L states,

$$|p_1|^2 + |q_1|^2 = |p_2|^2 + |q_2|^2 = 1\,, \tag{3}$$

the coefficients in Eqs. (1) are determined by six real parameters. Of these, three are arbitrary, reflecting the arbitrariness in the choice of an overall phase and the phases of $|S\rangle$ relative to $|L\rangle$, and of $|K^0\rangle$ relative to $|\bar{K}^0\rangle$, respectively. The remaining three are then defined with respect to the convention which fixes the arbitrary phases; we need not commit ourselves to any particular choice at this stage.

Inverting Eqs. (1), we have

$$|K^0\rangle = D(q_2 |S\rangle + q_1 |L\rangle)\,, \tag{4a}$$

$$|\bar{K}^0\rangle = D(p_2 |S\rangle - p_1 |L\rangle)\,, \tag{4b}$$

where

$$D^{-1} = (p_1 q_2 + p_2 q_1) \tag{5}$$

is tacitly assumed to be non-zero; otherwise, our initial premise would not apply. A state prepared initially as $|K^0\rangle$, for example in the reaction

$$\pi^- + p \to \Lambda^0 + K^0\,,$$

can be re-expressed in the form (4a) and, from the assumed timedependence Eq. (2) of S and L states, will evolve with time as

$$|K^0\rangle \to D(q_2 \theta_1 |S\rangle + q_1 \theta_2 |L\rangle)$$

or

$$|K^0\rangle \to D[(q_2 p_1 \theta_1 + q_1 p_2 \theta_2)|K^0\rangle + q_2 q_1 (\theta_1 - \theta_2)|\bar{K}^0\rangle] \tag{6}$$

re-introducing the definitions (1) of $|S\rangle$ and $|L\rangle$.

Similarly,

$$|\bar{K}^0\rangle \to D[p_1 p_2 (\theta_1 - \theta_2)|K^0\rangle + (p_2 q_1 \theta_1 + p_1 q_2 \theta_2)|\bar{K}^0\rangle]\,. \tag{7}$$

TCP-invariance requires

$$\langle \Theta\alpha | S | \Theta\beta \rangle = \langle \beta | S | \alpha \rangle \tag{8}$$

where $\Theta \equiv TCP$ represents the operation which reverses all helicities and replaces every particle by its corresponding antiparticle. Applying Eq. (8) to the case when both α and β are K^0 states, we see that TCP-invariance requires the K^0 amplitude from an initial K^0 state to equal the amplitude for an initial $\bar{K}^0$ state to remain a $\bar{K}^0$. Comparing Eqs. (6) and (7), this requires

$$q_2 p_1 = q_1 p_2 \tag{9}$$

or

$$q_1/p_1 = q_2/p_2 \tag{9a}$$

for TCP-invariance. Similarly, T-invariance requires [2]

$$\langle\tilde{\alpha}|S|\tilde{\beta}\rangle = \langle\beta|S|\alpha\rangle \tag{T 10}$$

where $|\tilde{\alpha}\rangle$ represents the state obtained from $|\alpha\rangle$ by reversing the momenta of all particles, keeping their helicities fixed. Choosing α, β to be K^0, $\bar{K}^0$ states respectively, T-invariance requires the amplitude for K^0 to transform into $\bar{K}^0$ to be the same as the amplitude for $\bar{K}^0$ to transform into K^0. From Eqs. (6) and (7), this requires

$$q_1 q_2 = p_1 p_2 \tag{T 11}$$

or

$$q_1/p_1 = p_2/q_2\,. \tag{T 11a}$$

The condition of CP-invariance

$$\langle CP\beta|S|CP\alpha\rangle = \langle\beta|S|\alpha\rangle \tag{12}$$

applied to the cases $\alpha = \beta = K^0$, and $\alpha = K^0, \beta = \bar{K}^0$, require both Eqs. (9) and (T 11) to apply. This is as expected since the simultaneous requirement of TCP- and T-invariance must be equivalent to the conditions imposed by CP-invariance. Eqs. (9) and (T 11) are the restrictions imposed on the structure of the S and L states by the requirements of TCP- and T-invariance respectively. A commonly used representation in the TCP-invariant case is

$$|S, L\rangle = [2(1+|\varepsilon|^2)]^{-\frac{1}{2}}\,[(1+\varepsilon)|K^0\rangle \pm (1-\varepsilon)|\bar{K}^0\rangle] \tag{9b}$$

which corresponds to the choice of L, S relative phase such that the amplitude of the K^0 component has the same phase in both.

Beside restricting the form of the S and L states, TCP- or T-invariance also imposes relations between decay amplitudes, which can be stated in the form of *Watson*'s theorem and its generalisation. We write the S-matrix as the sum of two parts S_0 and S_1, where S_0 is the form that the S-matrix would assume in the absence of interactions responsible for decay. We are interested in matrix-elements of S between a state $|c\rangle$

which would be stable in the absence of decay interactions, and other eigenstates $|\beta\rangle$ of S_0:

$$S_0 |\beta\rangle = e^{2i\delta_\beta} |\beta\rangle . \tag{13}$$

From the unitarity of S, we have

$$\langle c | S^+ S | \beta\rangle = \langle c | S_0^+ S_1 + S_1^+ S_0 | \beta\rangle = 0 \tag{14}$$

if we neglect terms of second order in S_1. Then, if we define $\boldsymbol{S} = \boldsymbol{I} + i\boldsymbol{T}$, it follows from the definitions of $|c\rangle$ and $|\beta\rangle$ that

$$\langle c | T | \beta\rangle = e^{2i\delta_\beta} \langle \beta | T | c\rangle^* . \tag{15}$$

Applying the condition (8) of TCP-invariance, we find

$$\langle \Theta\beta | T | \bar{K}^0\rangle = e^{2i\delta_\beta} \langle \beta | T | K^0\rangle^* . \tag{16}$$

It is clear from Eq. (8) that if $|\beta\rangle$ is a scattering eigenstate (13), then so is $|\Theta\beta\rangle$ with the same eigenphase δ_β. Eq. (16) is most easily satisfied by writing

$$\langle \beta | T | K^0\rangle = A_\beta e^{i\delta_\beta} , \tag{16a}$$

$$\langle \Theta\beta | T | \bar{K}^0\rangle = A_\beta^* e^{i\delta_\beta} . \tag{16b}$$

Similarly, the condition (T 10) of T-invariance applied to Eq. (15) leads to

$$\langle \beta | T | K^0, \bar{K}^0\rangle = e^{2i\delta_\beta} \langle \beta | T | K^0, \bar{K}^0\rangle^* , \tag{T 17}$$

which is satisfied by writing

$$\langle \beta | T | K^0\rangle = A_\beta e^{i\delta_\beta} , \tag{T 18a}$$

$$\langle \beta | T | \bar{K}^0\rangle = A'_\beta e^{i\delta_\beta} \tag{T 18b}$$

with A_β and A'_β *real.*

All consequences of TCP-invariance in neutral kaon decays follow from the relation (9) and (16); similarly, the requirements of T-invariance are contained in Eqs. (T 11) and (T 17).

An important consequence of TCP-invariance follows from a relation given by *Glashow* and *Weinberg* [3]. Consider a self-conjugate decay channel α which is also a scattering eigenstate. From Eqs. (9b) and (16a) and (16b), we have

$$\langle \alpha | T | S, L\rangle = [2(1 + |\varepsilon|^2)]^{-\frac{1}{2}} [(1+\varepsilon) A_\alpha \pm (1-\varepsilon)\lambda_\alpha A_\alpha^*] e^{i\delta_\alpha} . \tag{19}$$

where $\Theta|\alpha\rangle = \lambda_\alpha |\alpha\rangle$ with $\lambda_\alpha = \pm 1$.

Therefore

$$2\,\mathrm{Re} \langle \alpha | T | L\rangle^* \langle \alpha | T | S\rangle = \langle L | S\rangle (|\langle \alpha | T | S\rangle|^2 + |\langle \alpha | T | L\rangle|^2) . \tag{20}$$

For channels containing hadrons, one cannot in general directly measure decay amplitudes to scattering eigenstates $|\alpha\rangle$. For such cases, one obtains a useful relation by summing Eq. (20) over groups of scattering eigenstates composed of the same set of physical channels. If $\alpha_1\alpha_2\ldots\alpha_n$ represent the scattering eigenstates formed from the physical channels $j_1j_2\ldots j_n$, the unitarity of the transformation from j to α states assures that the summation of Eq. (20) over the set $\alpha_1\alpha_2\ldots\alpha_n$ is equivalent to summation over $j_1j_2\ldots j_n$. Thus

$$2\,\mathrm{Re}\sum_j \langle j|T|L\rangle^* \langle j|T|S\rangle = \langle L|S\rangle \sum_j (\gamma_1^j + \gamma_2^j)\,. \tag{21}$$

An analogous relation holds for the *T*-invariant case [4]. An important further constraint that should apply in general is the requirement of unitarity, which we state in the form given by *Bell* and *Steinberger*. For an arbitrary initial state

$$|\psi(0)\rangle = a_1\,|S\rangle + a_2\,|L\rangle \tag{22}$$

which evolves with time according to the prescription (2),

$$N(t) = \langle\psi(t)|\psi(t)\rangle = |a_1|^2\,\mathrm{e}^{-\gamma_1 t} + |a_2|^2\,\mathrm{e}^{-\gamma_2 t} + 2\,\mathrm{Re}\left[a_2^* a_1\,\mathrm{e}^{i(\lambda_2^* - \lambda_1)t}\langle L|S\rangle\right], \tag{23}$$

the rate of decrease of the norm

$$\left.\frac{-\mathrm{d}N}{\mathrm{d}t}\right|_{t=0} = \gamma_1|a_1|^2 + \gamma_2|a_2|^2 - 2\,\mathrm{Re}\left[i a_2^* a_1(\lambda_2^* - \lambda_1)\,\langle L|S\rangle\right] \tag{24}$$

must be accounted for by the rate of transitions out of the initial state (22):

$$\sum_k |\langle k|T|a_1 S + a_2 L\rangle|^2 \tag{25}$$

where T has been normalised to yield the transition rate directly. Since the equality of (24) and (25) must hold for every superposition of the form (22), we must equate coefficients of $|a_1|^2$, $|a_2|^2$ and $a_2^* a_1$ in (24) and (25). Thus [5]

$$\gamma_1 = \sum_k \gamma_1^k = \sum_k |\langle k|T|S\rangle|^2\,, \tag{26}$$

$$\gamma_2 = \sum_k |\langle k|T|L\rangle|^2\,, \tag{27}$$

$$\left[(m_2 - m_1) + \tfrac{1}{2}i(\gamma_1 + \gamma_2)\right]\langle L|S\rangle = i\sum_k \langle k|T|L\rangle^* \langle k|T|S\rangle\,. \tag{28}$$

The imaginary part of Eq. (28) is equivalent to Eq. (21) summed over *all* decay channels, in the *TCP*-invariant case. A corresponding relation holds in the case of *T*-invariance.

A Simple Model

Before proceeding to the analysis of $K^0 \to 2\pi$ decay, it will be instructive to consider a simple model discussed by *Weinberg* [6], which clearly illustrates the basic physical considerations. Suppose that K^0 and $\bar{K}^0$ states possessed only one self-conjugate decay channel, let us call it $\sigma^0\sigma^0$. Then, since TCP-invariance Eq. (16) requires

$$\langle\sigma^0\sigma^0|T|\bar{K}^0\rangle = e^{2i\delta_\sigma}\langle\sigma^0\sigma^0|T|K^0\rangle^*,$$

the decay amplitudes for K^0 and $\bar{K}^0$ are equal in magnitude and differ at most by a phase-factor, which can be eliminated by adjusting the phase of $\bar{K}^0$ relative to K^0. Then the superposition

$$|K_-\rangle = (|K^0\rangle - |\bar{K}^0\rangle)/\sqrt{2}$$

would have a vanishing decay amplitude. An arbitrary neutral kaon state would in general have a non-vanishing projection on this state $|K_-\rangle$ and one might be tempted to conclude that, in this simple model, TCP-invariance alone would lead almost every neutral kaon state to leave a residual stable component of the form $|K_-\rangle$, which would be incapable of decay. This is *not* true in general because the composition of the states $|S\rangle$ and $|L\rangle$, which decay exponentially *preserving their structure*, is determined not only by real energy-conserving decays but also by virtual transitions to intermediate states of higher or lower energy, which thereby induce $K^0 \leftrightarrow \bar{K}^0$ transformations. Let us write $|S\rangle$ and $|L\rangle$ in the form (9b) and adjust the phase of $|K^0\rangle$ and $|\bar{K}^0\rangle$ so that

$$\langle\sigma^0\sigma^0|T|K^0\rangle = \langle\sigma^0\sigma^0|T|\bar{K}^0\rangle = A_\sigma e^{i\delta_\sigma}. \tag{29}$$

Then, applying Eq. (19), we have

$$\langle\sigma^0\sigma^0|T|S\rangle = 2^{\frac{1}{2}}(1+|\varepsilon|^2)^{-\frac{1}{2}}A_\sigma e^{i\delta_\sigma} \tag{30}$$

and

$$\langle\sigma^0\sigma^0|T|L\rangle = \varepsilon\langle\sigma^0\sigma^0|T|S\rangle. \tag{31}$$

The physical origin of the last relation can be understood better by writing

$$|L\rangle = [2(1+|\varepsilon|^2)]^{-\frac{1}{2}}[(|K^0\rangle - |\bar{K}^0\rangle) + \varepsilon(|K^0\rangle + |\bar{K}^0\rangle)]\,; \tag{32}$$

the decay of the state $|L\rangle$ is determined completely by its

$$K_+ = (K^0 + \bar{K}^0)/\sqrt{2}$$

component. Only if some other consideration, e.g. CP-invariance, forbade the admixture of such a component with K_-, would we get a stable state. The unitarity relations (26) and (27) yield

$$\gamma_1 = 2(1+|\varepsilon|^2)^{-1}|A_\sigma|^2\,, \tag{33}$$

$$\gamma_2 = |\varepsilon|^2\gamma_1 \tag{34}$$

while Eq. (28) can be written as

$$\varepsilon = \tfrac{1}{2}(1+|\varepsilon|^2)\,[1+2i(m_2-m_1)/(\gamma_1+\gamma_2)]\,\langle S|L\rangle \tag{35}$$

for this simple model. With the phase-convention implied by Eq. (9b), $\langle S|L\rangle$ is a real number so that Eq. (35) fixes the phase of ε to be*

$$\phi_W = \tan^{-1}[2(m_2-m_1)/(\gamma_1+\gamma_2)] \tag{36}$$

and the single real parameter $|\varepsilon|$ determines all *CP*-nonconserving effects.

In the case when there are several decay channels α, if all decay amplitudes A_α are relatively real (i.e. apart from the effect, Eq. (16), of final-state interactions), the formulae (33)–(36) still apply* provided that $|A_\sigma|^2$ in Eq. (33) is replaced by $\sum_\alpha |A_\alpha|^2$. All amplitude-ratios

$$\eta_\alpha = \langle\alpha|T|L\rangle/\langle\alpha|T|S\rangle \tag{37}$$

are given by the same complex parameter ε, whose phase is fixed by Eq. (36). These are the characteristics of the "superweak" theory of *CP*-nonconservation, in which all $|\Delta S|\leqq 1$ interactions are *CP*-conserving so that *CP*-nonconserving transitions in neutral kaon decay are induced only by $K_-\leftrightarrow K_+$ transformations. The single parameter $|\varepsilon|$ contains all information about *CP*-nonconserving effects, which are restricted to the neutral kaon system if we neglect the extremely feeble (see below) $|\Delta S|=2$ decay amplitudes which the *CP*-noninvariant superweak interaction would induce, in principle.

$K^0\to 2\pi$ Decays

The model in which neutral kaons have only one decay mode is actually quite a good approximation to the real situation, where the amplitude for decay to the $I=0$ channel dominates all other amplitudes [7]. Accordingly, it is convenient to choose the phases of K^0 and $\bar{K}^0$ states so that [8]

$$\langle 2\pi(I=0)|T|\bar{K}^0\rangle = \langle 2\pi(I=0)|T|K^0\rangle = A_0\,e^{i\delta_0}$$

where A_0 is a real positive number, and δ_0 is the $I=0$ $\pi\pi$ *s*-wave phase shift at an energy equal to the K^0-mass. Then, for any other decay

* These statements hold strictly if $\lambda_\alpha=+1$ for all channels. The equations obtained by replacing γ_2 by $-\gamma_2$ in Eq. (36) and η_α by $(\eta_\alpha)^{\lambda_\alpha}$ in Eq. (37) remain valid when $\lambda_\alpha=-1$ channels also occur, if $|\varepsilon|^2$ can be neglected in comparison with unity.

channel represented by the scattering eigenstate β, the K^0 and $\bar{K}^0$ decay amplitudes will be given by Eqs. (16a) and (16b). The departure from reality of any decay amplitude A_β then indicates the CP-non-invariance in that amplitude *relative* to the $2\pi(I=0)$ decay amplitude. We can rewrite Eq. (19), for $\lambda_\alpha = 1$, as

$$\langle\alpha|T|S\rangle = 2^{\frac{1}{2}}(1+|\varepsilon|^2)^{-\frac{1}{2}}[\mathrm{Re}\,A_\alpha + i\varepsilon\,\mathrm{Im}\,A_\alpha]\,\mathrm{e}^{i\delta_\alpha},$$

$$\langle\alpha|T|L\rangle = 2^{\frac{1}{2}}(1+|\varepsilon|^2)^{-\frac{1}{2}}[i\,\mathrm{Im}\,A_\alpha + \varepsilon\,\mathrm{Re}\,A_\alpha]\,\mathrm{e}^{i\delta_\alpha}.$$

Then

$$\langle\pi^-\pi^+|T|S\rangle = 2^{\frac{1}{2}}(1+|\varepsilon|^2)^{-\frac{1}{2}}\left[\sqrt{\tfrac{2}{3}}A_0\,\mathrm{e}^{i\delta_0} + \sqrt{\tfrac{1}{3}}(\mathrm{Re}A_2 + i\varepsilon\,\mathrm{Im}A_2)\mathrm{e}^{i\delta_2}\right], \tag{38a}$$

$$\langle\pi^-\pi^+|T|L\rangle = 2^{\frac{1}{2}}(1+|\varepsilon|^2)^{-\frac{1}{2}}\left[\sqrt{\tfrac{2}{3}}\varepsilon A_0\,\mathrm{e}^{i\delta_0} + \sqrt{\tfrac{1}{3}}(i\,\mathrm{Im}A_2 + \varepsilon\,\mathrm{Re}A_2)\mathrm{e}^{i\delta_2}\right], \tag{38b}$$

where A_2 is the K^0 decay amplitude to the $I=2$ $\pi\pi$ state. We have neglected electromagnetic corrections of relative order α; such corrections will cause the scattering eigenstates to be changed slightly from isospin eigenstates, and the Clebsch-Gordan coefficients would be modified accordingly. To lowest order in the parameters ε and

$$\varepsilon' = i2^{-\frac{1}{2}}(\mathrm{Im}\,A_2)/A_0\,\mathrm{e}^{i(\delta_2-\delta_0)}, \tag{39}$$

we find

$$\eta_{+-} = \frac{\langle\pi^+\pi^-|T|L\rangle}{\langle\pi^+\pi^-|T|S\rangle} = \varepsilon + \varepsilon'/(1+\Delta) \tag{40}$$

where

$$\Delta = 2^{-\frac{1}{2}}(\mathrm{Re}\,A_2)/A_0 \cdot \mathrm{e}^{i(\delta_2-\delta_0)}.$$

Similarly

$$\eta_{00} = \frac{\langle\pi^0\pi^0|T|L\rangle}{\langle\pi^0\pi^0|T|S\rangle} = \varepsilon - 2\varepsilon'/(1-2\Delta). \tag{41}$$

The ε on the R.H.S. of Eqs. (40), (41) represents the contribution to the CP-nonconserving $L\to 2\pi$ transition amplitude from the CP-impurity in the state $|L\rangle$, which is therefore the same in all channels, while ε' measures the CP-nonconserving part of the $I=2$ $\pi\pi$ decay amplitude relative to the $I=0$ amplitude. The denominators on the R.H.S. of Eqs. (40) and (41) are usually omitted on the basis of the $\Delta I = 1/2$ rule; the smallness of Δ can be checked [9] if one knows the phase $(\delta_2-\delta_0)$ in addition to η_{+-}, η_{00}, and the charged/neutral ratio C in $K_s\to 2\pi$ decay, given by

$$C = \left|\frac{\langle\pi^+\pi^-|T|S\rangle}{\langle\pi^0\pi^0|T|S\rangle}\right|^2 = 2\left|\frac{1+\Delta}{1-2\Delta}\right|^2 \tag{42}$$

to lowest order in ε, ε'. We have heard in Prof. *Steinberger*'s lectures that the experimental situation is still rather uncertain. We therefore confine ourselves to some general remarks. Assuming TCP-invariance and the phase-convention corresponding to the definition (9b) of $|S\rangle$ and $|L\rangle$ states, $\langle S|L\rangle$ is a real number and the unitarity condition (28) requires that

$$\mathrm{Im}\left\{e^{-i\phi_W}\sum_k \eta_k \gamma_1^k\right\}=0 \tag{43}$$

where ϕ_W was defined in Eq. (36), η_k in Eq. (37), and γ_1^k is the partial decay rate from the state $|S\rangle$ into the channel k. Since 2π decay amplitudes are dominant, it is convenient to separate the contribution Z of these channels and to write

$$(\gamma_1^{2\pi})^{-1}\sum_k \eta_k \gamma_1^k = Z + \sum_{k\neq 2\pi}{}' \eta_k(\gamma_1^k/\gamma_1^{2\pi}) \tag{44}$$

where

$$Z=(C\eta_{+-}+\eta_{00})/(C+1)\,. \tag{45}$$

Contributions of other decay channels can be taken into account explicitly if the corresponding partial decay rates γ_1^k and partial amplitude ratios η_k are known. Where the complex parameters η_k are unknown, the partial decay rates γ_1^k and γ_2^k furnish the *magnitude* $(\gamma_1^k\gamma_2^k)^{\frac{1}{2}}$ of $\eta_k\gamma_1^k$ and thereby a limit on the absolute magnitude of the remaining sum on the R.H.S. of Eq. (44). Eq. (43) thus leads to the inequality

$$|\mathrm{Im}(e^{-i\phi_W}Z)| \leqq M_\Theta\,. \tag{46}$$

If some of the partial decay rates are not known, e.g. from the state $|S\rangle$ to 3π channels, the experimental upper limits provide corresponding upper bounds. A detailed discussion [4] based on available experimental information leads to a present limit of about 6×10^{-4} for M_Θ. If M_Θ could be neglected, Eq. (46) would require [9] the complex vector Z to have the phase ϕ_W and therefore to lie along the straight line $\mathscr{L}_\Theta$ through the origin making an angle $\phi_W \approx 43°$ with the positive x-axis. For a finite value of M_Θ, the condition (46) imposed by unitarity and TCP-invariance, relaxes to the requirement that Z lie within a distance M_Θ of $\mathscr{L}_\Theta$. The allowed region is indicated by the shaded strip in Fig. 1. The application of Eq. (21) to 2π channels yields another consequence of TCP-invariance [10],

$$\mathrm{Re}\,Z \approx \mathrm{Re}\,\varepsilon \approx \tfrac{1}{2}\langle S|L\rangle \tag{47}$$

where the parameter $\langle S|L\rangle$ is also required, by TCP-invariance, to be equal to the charge-asymmetry $\mathscr{A}_l$ in leptonic decays of the state $|L\rangle$, multiplied by a proportionality factor $C(x_l)$ (see Appendix A). If the

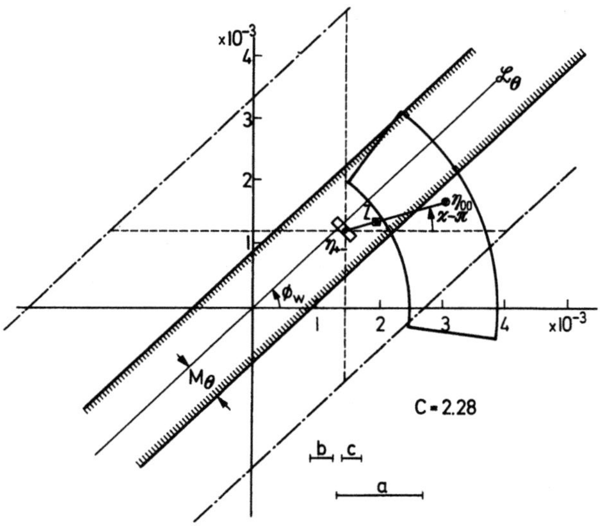

Fig. 1. The small rectangular box indicates the value of η_{+-}. Unitarity and TCP-invariance require Z to fall within the shaded strip, and the dot-dashed lines mark the corresponding boundaries within which η_{00} must lie, for the central value of η_{+-}. The large annular box represents the value of η_{00} reported by *Chollet et al.* (Ref. [12]). Values of Reε deduced from measurements of leptonic charge-asymmetry in K_L decays are indicated by *a*, *b*, and *c*

$\Delta S = \Delta Q$ rule is valid, this factor is exactly unity, whereas if $\Delta S = -\Delta Q$ transitions also occur, $C(x_l)$ can be deduced from measurements of the time-dependence of the corresponding leptonic decay mode of neutral kaons. We have heard from Prof. *Steinberger* that such measurements show that $C(x_l)$ is close to unity for $\pi e \nu$ decays. The measurement of the leptonic charge-asymmetry thus specifies the ordinate on which Z must lie.

Tests of *T*-Invariance

A convenient representation of S and L states under the hypothesis of T-invariance is

$$|S\rangle = [2(1+|\zeta|^2)]^{-\frac{1}{2}}\,[(1+\zeta)|K^0\rangle + (1-\zeta)|\bar{K}^0\rangle]\,, \qquad \text{(T 48a)}$$

$$|L\rangle = [2(1+|\zeta|^2)]^{-\frac{1}{2}}\,[(1-\zeta)|K^0\rangle - (1+\zeta)|\bar{K}^0\rangle]\,, \qquad \text{(T 48b)}$$

which obviously satisfies Eq. (T 11). Then $\langle S|L\rangle$ is purely imaginary and application of this condition to the unitarity relation (28) gives

$$\mathrm{Re}\left\{e^{-i\phi_W}\sum_k \eta_k \gamma_1^k\right\} = 0 \qquad \text{(T 49)}$$

leading to the inequality

$$|\mathrm{Re}(e^{-i\phi_W} Z)| \leqq M_T \qquad \text{(T 50)}$$

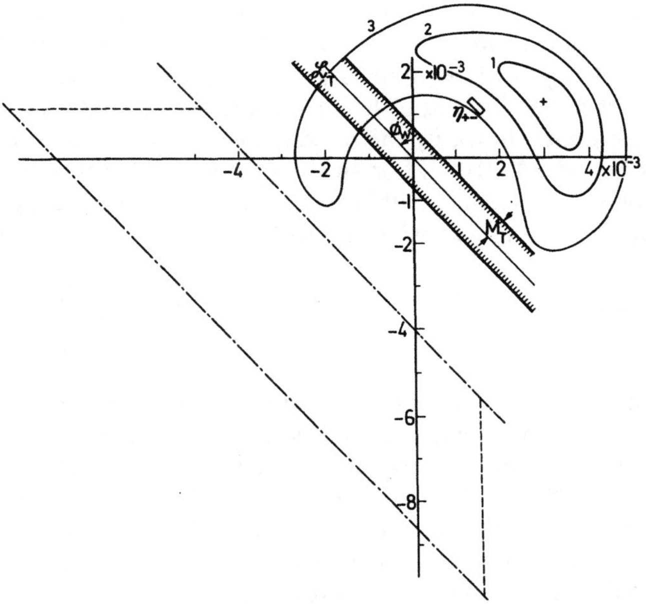

Fig. 2. Unitary and T-invariance require Z to fall within the shaded strip. For the central value of η_{+-}, this requires η_{00} to lie in the band enclosed by dot-dashed lines. The contours correspond to 1, 2, and 3 standard deviations from the central value of η_{00} (shown by the cross) as reported in Ref. [12]

if we proceed exactly as in the TCP-invariant case. Unitarity and T-invariance thus require Z to lie within a distance M_T of the straight line $\mathscr{L}_T$ drawn through the origin at an inclination of $(\pi/2+\phi_W)\approx 133°$ to the positive real axis [4]. The present limit on M_T is about 5×10^{-4}; the allowed strip corresponding to this value is shown by shaded boundaries in Fig. 2. For the value of η_{+-} quoted by Prof. *Steinberger* at the CERN Conference in January [11], shown by the small rectangular box, the condition (T 50) imposed by unitarity and T-invariance, requires η_{00} to lie within the band parallel to $\mathscr{L}_T$ bounded by dot-dashed lines. This requires the magnitude of η_{00} to exceed a minimum value, of approximately 2.8×10^{-3}, and to have a phase essentially opposite to that of η_{+-}. The reported measurement [12] of ϕ_{00} in the first quadrant therefore disproves the hypothesis of T-invariance if we assume unitarity. (The possible slight difference in the relative phase of S and L states as defined in Eqs. (9b) and (T 48) could in principle cause the same experimental result to be represented by η_k's with slightly different phases in the two cases. Since both ε and ζ can be shown to be of modulus less than 10^{-2}, this has no practical effect on our conclusions.) In view of the doubts surrounding the value of η_{00}, it is interesting to consider another argument against T-invariance which requires no knowledge of η_{00} except that it is not too large [13]. We have already explained that all values of η_{00} satisfying unitarity and T-invariance must fall within the dot-dashed band in Fig. 2; let us suppose further that it does not lie outside the limits shown by the broken lines – this will be true if $|\eta_{00}|$ does not exceed 5.4×10^{-3}. For all values of η_{00} within these limits, the phase χ of $(\eta_{+-} - \eta_{00})$ lies within the first quadrant. The $\pi\pi$ phase-shift combination $\psi = (\delta_2 - \delta_0)$, which is given, assuming T-invariance, by

$$\tan\psi = \tan\chi \left[\frac{(2C-1)(C-2)\tan^2\chi + 2(C+1)^2}{9C} \right]^{-\frac{1}{2}} \qquad \text{(T51)}$$

if we assume the $\Delta I = \frac{1}{2}$ rule, $|\Delta| \ll 1$, cannot then be reconciled with the *negative* values of $\tan\psi$ reported from measurements of $\pi\pi$ scattering [14].

Thus, unitarity and T-invariance cannot be satisfied for any phase ϕ_{00} if $|\eta_{00}| < 2.8 \times 10^{-3}$. For $2.8 \times 10^{-3} < |\eta_{00}| < 5.4 \times 10^{-3}$, ϕ_{00} must lie in or near the third quadrant for these conditions to be satisfied. In that case, the value of $(\delta_2 - \delta_0)$ deduced assuming the $\Delta I = \frac{1}{2}$ rule disagrees with the values reported from $\pi\pi$ scattering.

The Dynamical Origin of CP-Noninvariance

Once we accept the experimental evidence that CP-invariance is not a universal symmetry, it becomes logically impossible to locate the source of CP-noninvariance, just as it is impossible to decide whether the departure from sphericity of the Earth's figure is due to flattening of the poles or to bulging of the equator. In the latter case, we are dealing with two different descriptions of the same phenomenon; in the same way, the attribution of the property of CP-noninvariance to a particular type of interaction may well depend on the mode of description we choose to employ. Keeping this in mind, it is convenient for the purpose of analysis to assume that there is a specific type of interaction, named the X-interaction by Okun, which is responsible for CP-nonconservation in neutral kaon decays. That is to say, we assume that if the corresponding term $\mathscr{H}_X$ could be removed from the total Hamiltonian $\mathscr{H}$ describing elementary particle interactions, the remaining Hamiltonian $\tilde{\mathscr{H}}$ would be CP-invariant and all observable phenomena would satisfy this symmetry. This has been the traditional approach to broken symmetry in physics, which does not however guarantee that the method will succeed in the present instance. Nevertheless, we shall see that it has already led to many very interesting suggestions; and that a large variety of X-interactions have been proposed to account for CP-nonconservation in K^0 decays.

Superweak Theories

We have already discussed the situation in which CP-nonconservation in neutral kaon decay occurs only through the admixture ε of a CP-even component $K_+ = 2^{-\frac{1}{2}}(K^0 + \bar{K}^0)$ into the long-lived state $|L\rangle$, Eq. (32). We have also seen that TCP-invariance requires the amplitude for a K^0 to remain a K^0 to equal the corresponding amplitude for a $\bar{K}^0$ to remain $\bar{K}^0$, thus strangeness-conserving TCP-invariant interactions cannot develop a K_+ component from a state prepared initially as $|K_-\rangle = (|K^0\rangle - |\bar{K}^0\rangle)/\sqrt{2}$. Such an admixture can arise only from $K^0 \to \bar{K}^0$ and $\bar{K}^0 \to K^0$ transitions, which are *not* required to have equal amplitudes unless T-invariance or CP-invariance applies. Therefore, assuming TCP-invariance, a non-vanishing value of ε in Eq. (32) arises from CP-noninvariant interactions obeying $|\Delta S| = 2$. If ε is the sole indicator of CP-noninvariance, this may be because the primary CP-noninvariant interaction is a nonleptonic interaction obeying $|\Delta S| = 2$ which can contribute to $K^0 \leftrightarrow \bar{K}^0$ transitions and therefore to ε but not to any K^0 decay amplitude. In that case, we can estimate the required strength of the X_2-interaction from the admixture of K_+ which it introduces into the state $|L\rangle$, using perturbation theory,

$$\varepsilon = \frac{\langle K_+ | \mathscr{H}_{X_2} | K_- \rangle}{(\lambda_2 - \lambda_1)} . \tag{52}$$

Since $\eta_{+-} = \varepsilon$ in this theory, we require

$$\langle K_+ | \mathscr{H}_{X_2} | K_- \rangle \sim 10^{-3} (\lambda_2 - \lambda_1) .$$

If we characterise $\mathscr{H}_{X_2}$ by a coupling constant F_2 similar to the Fermi coupling constant G, we require

$$F_2 \sim 10^{-3} G^2 m_p^2 \sim 10^{-8} G \tag{53}$$

since $(\lambda_2 - \lambda_1)$ is a quantity of second order in G; the factor m_p^2 has been introduced to give F_2 the same dimensionality as G. We therefore see that an extremely weak X_2-interaction [15], weaker than second order weak interactions, suffices to explain CP-nonconservation in K^0 decay *if* all CP-noninvariant phenomena can be described by a single parameter ε with the phase ϕ_W, Eq. (36). Since the $K_L - K_S$ mass-difference is the only known manifestation of higher-order weak interactions, such a "superweak" X_2-interaction would not have any other observable effect. In contrast to this genuinely superweak interaction, another situation is possible where a single parameter ε with the phase ϕ_W would account for all CP-noninvariant phenomena in K^0-decay; this would be if CP-noninvariance resided in X-interactions which contribute negligibly to K^0 and $\bar{K}^0$ decay amplitudes. In that case also, all decay amplitudes would be CP-conserving but $K^0 \leftrightarrow \bar{K}^0$ transitions could occur through virtual intermediate states where the X-interaction would operate. If this were so, CP-noninvariant phenomena would not be confined to neutral kaon decay and one could expect to find much larger effects elsewhere under suitable circumstances. An example of such a theory would be if $\Delta S = -\Delta Q$ transitions were maximally CP-noninvariant [16] and increased in strength relative to $\Delta S = \Delta Q$ amplitudes from a negligible value at K^0 decay energies to approximate equality at higher energies. In that case, of course, one should expect large CP-noninvariant effects and large departures from the $\Delta S = \Delta Q$ rule in neutrino experiments at correspondingly high energies. As long as we confine our attention to neutral kaon decays, these theories are indistinguishable from the genuine $|\Delta S| = 2$ superweak theory, characterised by the coupling strength (53), so they may be described as "virtual superweak" theories, and we shall refer to them again later.

Milliweak Theories

The most straightforward interpretation of $K_L \to 2\pi$ decays would be that the $K^0 \to 2\pi$ decay interaction itself contains a CP-noninvariant part [8]. If we take the ratio of $K_L \to 2\pi$ and $K_S \to 2\pi$ decay amplitudes as representative of the strength of this X_1-interaction, obeying $|\Delta S| = 1$, relative to the usual weak interactions, it must be characterised by a coupling constant

$$F_1 \sim 10^{-3} G. \tag{54}$$

Such a "milliweak" interaction would cause CP-noninvariant amplitudes with relative magnitude 10^{-3} to be present in other non-leptonic $|\Delta S| = 1$ transitions, and although such effects have been strenuously sought ever since the discovery of parity nonconservation, no positive effect has yet been reported. Unfortunately, a precision exceeding 0.1 percent has not been attained in any of the experiments, so that an interaction with the strength (54) could be present without having been detected so far.

Within the context of the current-current interaction theory of weak interactions, such an X_1-interaction could arise in two different ways. Either the usual charged currents can be modified so as to introduce CP-noninvariance, or additional currents may be brought in, which are responsible for CP-nonconservation. A theory of the latter type, proposed by *Oakes* [17], introduces CP-noninvariance through the neutral octet currents j_λ^6 and j_λ^7 which were not physically identified, except in terms of commutators of other known currents, in Gell-Mann's $SU(3) \times SU(3)$ algebra [18]; it will be discussed in the seminar by Dr. *Gronau*. CP-noninvariance was incorporated into the charged current-current interaction theory by *Glashow* [19], by arbitrarily introducing phase-differences between the vector and axial-vector components of the weak current. Since no indication of T-noninvariance has been seen in nuclear β-decay experiments, the relative phase φ between the $\Delta S=0$ components must be quite small, $|\varphi|<10^{-2}$, and was set equal to zero; since experimental information on hyperon β-decays is much scarcer, the relative phase Φ between the $\Delta S=1$ currents could be taken to be much larger without contradicting observation. By applying *Gell-Mann*'s current algebra and PCAC to the usual current-current interaction theory (with $\Phi=\varphi=0$), several authors [20] showed that nonleptonic decays of K-mesons would obey the $\Delta I=\frac{1}{2}$ rule even though the current-current interaction in general contains both $\Delta I=\frac{1}{2}$ and $\Delta I=\frac{3}{2}$ parts. If CP-invariance is not assumed, the argument no longer applies and the generalisation [21] is that the $\Delta I=\frac{1}{2}$ rule is obeyed by the CP-invariant part of the interaction if $\Phi=\pm\varphi$, and by the CP-noninvariant part if $\Phi=-\varphi$. The choice $\Phi=\varphi$ would make the CP-invariant interaction obey the $\Delta I=\frac{1}{2}$ rule and introduce a CP-noninvariant part obeying $\Delta I=\frac{3}{2}$ with relative strength of order φ, as would be required if $\eta_{+-}\neq\eta_{00}$. Since φ is known to be a small angle, from nuclear β-decay, this would require CP-noninvariant effects in all $|\Delta S|=1$ interactions, leptonic or non-leptonic, to be of the same order of smallness $\varphi\sim10^{-3}$. On the other hand, the choice $\Phi=-\varphi$ would preserve this conclusion, and also assure that $\eta_{+-}=\eta_{00}$. The smallness of φ is not, however, explained by the theory but chosen to fit the observations.

Millistrong Theories

Another interesting possibility is that the interaction responsible for CP-nonconservation in K^0 decays conserves strangeness [22], $\Delta S=0$. Such an X_0-interaction could induce CP-nonconserving amplitudes in K^0 decay through its joint action with the usual CP-conserving non-leptonic weak interaction obeying $|\Delta S|=1$. For example, the CP-nonconserving amplitude for $K_-\to2\pi$ decay could be understood as occurring through processes like

$$K_-\xrightarrow{|\Delta S|=1}A\xrightarrow{X_0}2\pi \tag{55}$$

where A is a neutral meson system with $S=0$, $CP=-1$. The virtual transition from K_- to A occurs through the action of the CP-conserving $|\Delta S|=1$ weak interaction, while the transition from A to the final 2π state is caused by the X_0-interaction. If the strength of this interaction is represented by the dimensionless coupling constant f_0, the amplitude for the process (55) will be of order Gf_0. Since we require it to be of the order of 10^{-3} of the "normal" weak amplitude for $K_+ \to 2\pi$ decays, of order G, we require

$$f_0 \sim 10^{-3}. \tag{56}$$

In this case, the X-interaction must be much *stronger* than weak interactions, in fact comparable in strength to electromagnetic interaction. Since, by definition, the X-interaction violates CP-invariance, and P-invariance has been checked to an accuracy of much better than a part in 10^3, as we shall hear in detail from Prof. *Telegdi*, it is necessary to postulate that such an X_0-interaction must be essentially parity-conserving and therefore violate C-invariance. By the TCP theorem, this would imply a corresponding degree of T-noninvariance. Recently, tests of T-invariance have been made in nuclear physics with a precision approaching the required level, but no positive indication of T-violation has been published so far. Again, as in the case of milliweak CP-nonconserving interactions, one requires in general experiments with an accuracy of better than 0.1 percent in order to detect the effect of such interactions. *Friedberg*, *Lee*, and *Schwartz* noted that η^0 decays, which occur through interactions of strength comparable to electromagnetic interactions, provide an excellent opportunity for the X_0-interaction to produce relatively large C-noninvariant effects. The most recent experiment [23] (with the largest statistics of 40000 events) shows that in $\eta^0 \to \pi^+\pi^-\pi^0$ decays, there is a charge-asymmetry (π^+'s being more energetic on average than π^-'s) of

$$\mathcal{A}_\eta = (1.5 \pm 0.5)\,\%. \tag{57}$$

Should this result be confirmed, the existence of C-noninvariant interactions much stronger than weak interactions would be directly established.

T. D. Lee and collaborators [24] also made the interesting observation that there was little experimental evidence to support the generally assumed C-invariance of hadron electromagnetic interactions. They therefore suggested that the electromagnetic interaction of hadrons might show large departures from C-invariance, in which case the X_0-interaction could just be the effect of second-order radiative corrections, arising from the virtual emission and reabsorption of photons. The smallness of the CP-nonconserving amplitudes in K^0 decay then reflects the smallness of the parameter α governing the strength of electromagnetic interactions. This theory, and its derivatives, are the only ones in which the magnitude of CP-nonconserving effects in K^0

decay is theoretically "predicted", in order of magnitude, rather than being adjusted to fit the observations. A possible difficulty for this theory is the very low limit already set on the electric dipole moment d_n of the neutron; the best value [25] being $e.2 \times 10^{-23}$ cm. If the electromagnetic interaction of hadrons itself violates T-invariance, the dipole moment of the neutron should be given directly by the parity-admixture in the neutron wavefunction, of order G, giving

$$d_n \sim \mathrm{e}.(Gm_p^2) \cdot \xi \cdot 1/m_p\,, \tag{58}$$

which yields a numerical estimate of order $e.10^{-21}$ cm, even allowing for a depression-factor $\xi = 10^{-2}$, in analogy with the reduction below (Gm_p^2) of the parity-admixture in nuclei. If the theory of C-noninvariant electromagnetic interaction for hadrons is not to disagree with experiment, it must contain a mechanism to suppress the neutron dipole moment.

a-particles and Related Chimera

If hadron electromagnetic interactions are not C-invariant, the hadronic current must contain $\mathscr{C}$-even parts in addition to the generally accepted $\mathscr{C}$-odd part, where $\mathscr{C}$ is now defined as the symmetry with respect to particle-antiparticle exchange which has been checked to an accuracy of one percent in the strong interactions of hadrons. *T. D. Lee* has proposed [26] that the existence of $\mathscr{C}$-even currents would be most easily accounted for if there existed new types of particles, called a-particles, whose electric charge is unchanged under the operation of $\mathscr{C}$. The effect of a-particles can be easily understood by considering a model given by *T. D. Lee*, which we describe in conventional language following *Okun* [27]. Let us suppose that there exists a strong C-noninvariant interaction between normal hadrons and the hypothetical a-particles

$$\mathscr{H}_{na} = f_a j_n J_a \tag{59}$$

where j_n is a current composed of hadron operators only and J_a is a similar current for a-particles; f_a is a coupling constant of order unity. Let us assume further that the current j_n is even with respect to the charge-conjugation operator $\mathscr{C}$ for normal hadrons while J_a is *odd* with respect to the charge-conjugation operator $\mathscr{C}_a$ which acts on a-particles only. Then $\mathscr{H}_{na}$ is *odd* with respect to overall charge-conjugation $C = \mathscr{C} \otimes \mathscr{C}_a$, and reactions involving both normal and a-particles would in general show large charge-asymmetries in the distribution of a-particles. However, since $\mathscr{H}_{na}$ is even with respect to $\mathscr{C}$, there should be no C-noninvariant effects in processes where all participating particles are normal hadrons. If the a-particles couple in the usual way to the electromagnetic field, e.g. $\bar{a}\gamma^\lambda a A_\lambda$, their electromagnetic interaction cannot be $\mathscr{C}$-invariant at the same time as the electromagnetic coupling of the normal particles. As a consequence, the electromagnetic interaction of normal hadrons acquires a $\mathscr{C}$-noninvariant part, since photons could be emitted or absorbed by virtual a-particles created and subsequently reabsorbed, by the interaction (59). Processes involving normal hadrons only, with no external photons, would also show $\mathscr{C}$-noninvariant effects, of relative order α, as a result of virtual photon exchange between normal hadrons and virtual a-particles. This could then provide a dynamical basis for the origin and coupling strength (56) of the X_0-interaction. Virtual a-particles may then be regarded as catalytic agents which induce $K_L \to 2\pi$ decay. Since no strongly interacting particles have so far been observed which show large charge-asymmetries, we must conclude that a-particles must be more massive than 2–3 GeV. Further, since strong interactions are $\mathscr{C}$-invariant, strong and minimal electromagnetic interactions conserve the charge of a-particles separately from that of normal particles. Therefore, a-particles cannot be emitted or absorbed singly through strong interactions, but are subject to an additive conservation law similar to strangeness.

Okun has considered a variant of the a-particle theory in which the a-particles, which are coupled to hadrons according to (59) as before, are assumed to be electrically neutral [27]. CP-nonconservation in K^0 decays can arise from these neutral a-particles if they possess an additional CP-*invariant* interaction with normal hadrons. The strength of this extra interaction must be so adjusted that its joint effect with the interaction (59) yields the X-interaction of the desired strength. For example, if the X-interaction obeys $\Delta S = 0$, its strength must be given by a coupling constant of order 10^{-3} so as to obtain an effective X_0-interaction with this coupling strength (56). An effective superweak X_2-interaction may be similarly obtained as a consequence of a $|\Delta S| = 2$ CP-invariant coupling of neutral a-particles to hadrons, with a coupling constant of the order of $10^{-13} - 10^{-15}$. Thus every type of X-interaction can be simulated by the catalytic action of such a-particles. Note, however, that the magnitude of CP-nonconservation in K^0 decay has to be inserted in all these cases.

An interesting suggestion made by *Gell-Mann* [28] was to identify a-particles with the intermediate bosons $W^{\pm}$ which are believed to transmit the weak interactions. However, *Kobzarev*, and *Okun* [29] noted that if $a^{\pm} \equiv W^{\pm}$ particles interact strongly with normal hadrons, the degree of CP-nonconservation must be the same in electromagnetic and weak interactions. Consider, for example, a hadron vertex in which a W^+ is emitted through its postulated "semi-weak" coupling

$$g = (GM^2/\sqrt{2})^{\frac{1}{2}} \tag{60}$$

with the usual weak current. This vertex will suffer corrections of relative magnitude f_a from processes in which a $W^+ W^-$ pair is emitted through the strong interaction (59) and the W^- is subsequently reabsorbed through the "semi-weak" interaction. Thus, *CP*-noninvariance will occur in the weak interactions in relative order f_a, without the intervention of virtual photons, and the hypothesis $a^{\pm} \equiv W^{\pm}$ can be reconciled with experience only if $f_a \sim 10^{-3}$. The magnitude of *CP*-nonconservation in K^0 decays would again remain unexplained.

A variation on this theme, suggested by *Gell-Mann* [30], which avoids this difficulty is to postulate that *W*'s participate in strong *CP*-noninvariant interactions *with each other*, and possibly with other undiscovered particles h', but not with normal hadrons. Then there would be no perceptible *CP*-noninvariance in strong or electromagnetic interactions of normal particles, but a milliweak *CP*-noninvariant interaction is induced by electromagnetic corrections to the weak interaction, of relative order α, arising from the exchange of a virtual photon between a virtual W (or a h') and a normal particle. Purely leptonic weak interactions should then also show *CP*-noninvariant effects of order α. $\mathscr{C}$-noninvariance does not appear in electromagnetic interactions of normal hadrons because the non-invariant electromagnetic interaction cannot operate without the virtual production of a W. However, the electric dipole moment of the neutron in this theory is in general of the same order of magnitude Ge as in the theory of electromagnetic $\mathscr{C}$-violation; a non-vanishing dipole moment requires parity nonconservation in addition to T-noninvariance, therefore a weak vertex must be included even when hadron electromagnetic interactions are $\mathscr{C}$-noninvariant. To avoid contradiction with the increasingly stringent limits on d_n, it may be necessary to impose additional conditions to suppress the dipole moment. According to *Gell-Mann*, there is no contribution of order Ge if all strong interactions of W are parity-conserving, and the W fields obey the condition $\partial_\lambda W_\lambda = 0$. For further speculations along this direction, the reader is referred to the original source [30].

Summary

We conclude that a large variety of dynamical theories can be proposed to account for *CP*-noninvariance found in K^0 decay, and the exploration of these various possibilities will provide employment for many physicists for a considerable time. It is also quite possible that such endeavours will uncover further surprises. The decisions which *can* be reached from K^0 decay measurements are shown in the logical diagram (Fig. 3). The origin of *CP*-noninvariance is still undetermined,

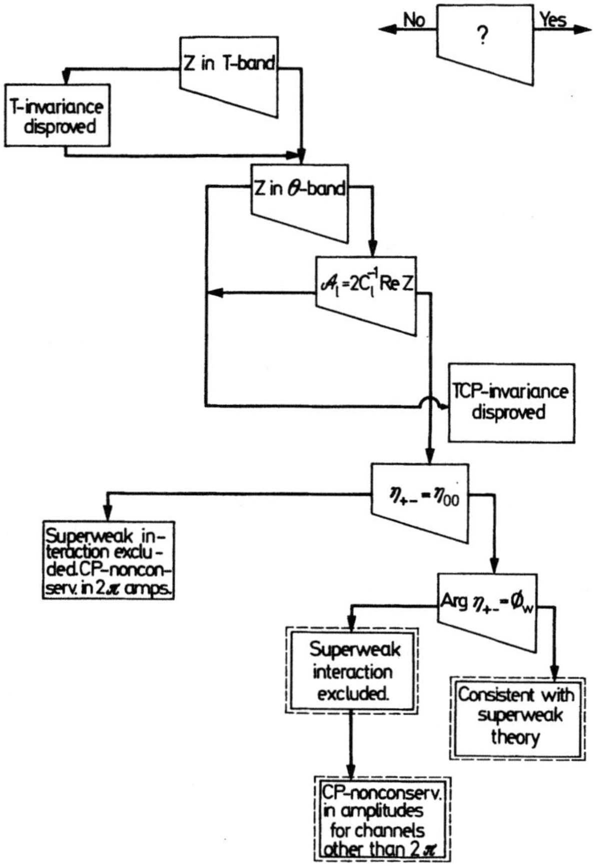

Fig. 3. Logical diagram of questions which can be decided from measurements on K^0 decay. Conclusions surrounded by broken lines depend on the additional hypothesis of TCP-invariance

although measurements may exclude certain possibilities. However, it is clear that as long as the only certain evidence of CP-noninvariance comes from K^0 decay, we can only have a partial view of the whole picture.

Appendix A

We illustrate how TCP-invariance can be systematically tested from measurements of neutral kaon decay rates, by considering decays from K^0 and $\bar{K}^0$ states to self-conjugate decay channels which are scattering eigenstates of the strong interactions (which automatically includes the case when interactions between particles in a decay channel can be neglected) [31].

From the hypothesis of exponential decay, Eq. (2), and the principle of superposition for decay amplitudes, the decay rate to *any* final state k is given by

$$R_{\pm}^{k}(t) = a_{\pm}^{k}\,\mathrm{e}^{-\gamma_1 t} + b_{\pm}^{k}\,\mathrm{e}^{-\gamma_2 t} + 2\,\mathrm{Re}\,[c_{\pm}^{k}\,\mathrm{e}^{i(\lambda_2^* - \lambda_1)t}] \tag{A 1}$$

where the subscripts $\pm$ refer to an initial K^0 and $\bar{K}^0$ state, respectively. For either case, the two real coefficients a^k, b^k and the complex coefficient c^k represent 4 real parameters which can be measured for every decay channel k. The exponential coefficients γ_1, γ_2 and $(m_2 - m_1)$ represent three more, so that there are $(8N+3)$ real quantities determinable from observations of all N K^0 and $\bar{K}^0$ partial decay rates.

The $2N$ conditions

$$a_{\pm}^{k}\, b_{\pm}^{k} = |c_{\pm}^{k}|^2 \tag{A 2}$$

are a consequence of the superposition principle alone. The conditions (9) imposed by TCP-invariance on the structure of S and L states imposes the further $3N$ conditions (at the cost of one unknown parameter)

$$\frac{a_{+}^{k}}{a_{-}^{k}} = \frac{b_{+}^{k}}{b_{-}^{k}} = \frac{-c_{+}^{k}}{c_{-}^{k}} = \left|\frac{1-\varepsilon}{1+\varepsilon}\right|^2 = \frac{1 - \langle S|L\rangle}{1 + \langle S|L\rangle} \tag{A 3}$$

and also apply to *any* final state k. For final states which are scattering eigenstates, the TCP conditions (16) on decay amplitudes yield the further N relations, equivalent to Eq. (20),

$$2\,\mathrm{Re}\, c_{+}^{\alpha} = (a_{+}^{\alpha} + b_{+}^{\alpha})\,\langle L|S\rangle\,. \tag{A 4}$$

The unitarity relations (26)–(28) provide three further conditions – the imaginary part of Eq. (28) is already contained in Eq. (A 4) – so that we have $(6N+2)$ independent relations between the $(8N+3)$ observable parameters in Eq. (A 1). Therefore $(2N+1)$ independent real parameters describe all decay rates from K^0 or $\bar{K}^0$ states. Since these must be determined by N independent complex decay amplitudes for the N channels and the complex structure-parameter ε, and the overall phase cannot be measured, Eqs. (A 3) and (A 4) exhaust the relations which can be written down as a consequence of TCP-invariance. It is left to the reader to verify that nothing further can be learnt by considering initial states which are arbitrary superpositions of K^0 and $\bar{K}^0$ states.

An important relation which follows from the application of this method of analysis to non-self-conjugate channels is

$$\mathscr{A}_{\alpha} = \frac{\gamma_2^{\alpha} - \gamma_2^{\bar{\alpha}}}{\gamma_2^{\alpha} + \gamma_2^{\bar{\alpha}}} = \frac{b_{\pm}^{\alpha} - b_{\pm}^{\bar{\alpha}}}{b_{\pm}^{\alpha} + b_{\pm}^{\bar{\alpha}}} = \frac{2\,\mathrm{Re}(c_{+}^{\alpha} - c_{+}^{\bar{\alpha}})}{(a_{+}^{\alpha} + a_{+}^{\bar{\alpha}} + b_{+}^{\alpha} + b_{+}^{\bar{\alpha}})}\langle L|S\rangle \tag{A 5}$$

where $\overline{\alpha}$ is the channel conjugate to α: $|\overline{\alpha}\rangle = |\Theta\alpha\rangle$. If α is taken as $(\pi^- e^+ \nu)$ or $(\pi^- \mu^+ \nu)$, this gives the leptonic charge-asymmetry formula used by Prof. *Steinberger*.

References and Footnotes

1. Our description follows *Bell, J. S., Steinberger, J.*, Proc. Oxford Intl. Conf. on Elementary Particles 1965 (Rutherford Laboratory, Chilton, 1966).
2. Equations which are valid only under the assumption of *T*-invariance are identified by the letter *T* in the equation-number.
3. *Glashow, S. L., Weinberg, S.*: Phys. Rev. Letters **14**, 835 (1965).
4. *Kabir, P. K.*: Nature **220**, 1310 (1968).
5. The derivation of these relations from *S*-matrix unitarity will be given in Professor Wick's lectures.
6. *Weinberg, S.*: Phys. Rev. **110**, 782 (1968).
7. This statement is based on the dominance of 2π channels and the assumed validity of the $\Delta I = \frac{1}{2}$ rule.
8. *Wu, T. T., Yang, C. N.*: Phys. Rev. Letters **13**, 380 (1964).
9. *Gourdin, M.*: Nucl. Phys. B **3**, 207 (1967); — *Field, G., Kabir, P.*: Z. Phys. **208**, 60 (1968).
10. *Patnaik, B. K., Kabir, P. K.*: Z. Physik **209**, 157 (1968).
11. Proc. CERN Topical Conf. on Weak Interactions (CERN, Geneva, 1969).
12. *Chollet, J.*: Proc. CERN Topical Conf. on Weak Interactions (CERN, Geneva, 1969), p. 309.
13. *Ashkin, J., Kabir, P. K.*: Phys. Rev., to be published. Related remarks were made earlier by *Casella, R. C.*: Phys. Rev. Letters **21**, 1128 (1968); **22**, 554 (1969).
14. *Marateck, S.*: Phys. Rev. Letters **21**, 1613 (1968).
15. *Wolfenstein, L.*: Phys. Rev. Letters **13**, 562 (1964).
16. *Sachs, R. G.*: Phys. Rev. Letters **13**, 286 (1964).
17. *Oakes, R. J.*: Phys. Rev. Letters **20**, 1539 (1968).
18. *Gell-Mann, M.*: Phys. Rev. **125**, 1067 (1962).
19. *Glashow, S. L.*: Phys. Rev. Letters **14**, 35 (1965).
20. *Suzuki, M.*: Phys. Rev. **144**, 1154 (1966). — *Hara, Y., Nambu, Y.*: Phys. Rev. Letters **16**, 875 (1966). — *Elias, D. K., Taylor, J. C.*: Nuovo Cimento **44**, 518 (1966).
21. *Pati, J. C.*: Phys. Rev. Letters **20**, 812 (1968). — *Holstein, B.*: Phys. Rev. **171**, 1668 (1968).
22. *Okun, L.*: Soviet J. Nucl. Phys. **1**, 938 (1965).
23. *Gormley, M.*: Phys. Rev. Letters **21**, 402 (1968); **22**, 108 (1969).
24. *Bernstein, J., Feinberg, G., Lee, T. D.*: Phys. Rev. **139**, B 1650 (1965).
25. *Miller, P. D.*: Phys. Rev. Letters **19**, 381 (1967).
26. *Lee, T. D.*: Phys. Rev. **140**, B 959 (1965).
27. *Okun, L.*: Soviet J. Nucl. Phys. **3**, 189 (1966).
28. Proc. XIII Intl. Conf. High Energy Physics 1966 Berkeley: Univ. Calif. Press 1967.
29. *Kobzarev, I. Yu., Okun, L.*: Phys. Letters B **27**, 172 (1968).
30. *Gell-Mann, M.*: Report to Coral Gables Conf. 1969, to be published.
31. A discussion including non-conjugate channels and channels which are not scattering eigenstates will appear in a joint paper with *G. V. Dass* and *B. G. Kenny*.

Dr. *P. K. Kabir*
Science Research Council
Rutherford High Energy Laboratory
Chilton, Didcot, Berkshire/England

Relations for Semileptonic Weak Interactions Involving Photons

W. KUMMER

Contents

Abstract 113
1. Introduction 113
2. General Considerations 115
3. Vector Current in $\pi \to e\nu\gamma$ 117
4. The Vector Current in $\pi^+ \to \pi^0 e\nu\gamma$ 118
References and Footnotes 124

Abstract

By means of a low energy theorem type consideration a description of electromagnetic symmetry breaking for the vector parts in those semileptonic weak interactions can be given, where also a photon is emitted. The examples $\pi \to e\nu\gamma$ and $\pi^+ \to \pi^0 e\nu\gamma$ are discussed.

I. Introduction

The increasing intensity of particle beams either from meson factories or from secondary beams of improved high energy accelerators will enable soon the observation of weak reactions in which also one or even more external photons participate. The larger part of such reactions which have been seen already is unfortunately of the nonleptonic type [1]

$$K^\pm \to \pi^\pm \pi^0 \gamma\,,$$
$$\Sigma^+ \to p\gamma\,,$$
$$\Sigma^+ \to n\pi\gamma\,,$$
$$K^\pm \to \pi^\pm \pi^+ \pi^- \gamma$$

and the theoretician faces thus the usual difficulties of weak nonleptonic reactions.

On the other hand, an additional photon in semileptonic reactions, where the current picture is reasonably well understood, introduces just one further (electromagnetic) current so that the rather powerful machinery of symmetries, current algebra etc. is readily applicable. Prominent examples are here the decays [1]

$$\pi \to e\nu\gamma\,,$$
$$K^+ \to \pi^0 e\nu\gamma$$

and the radiative μ-capture [2]

$$\mu^- + p \to n + \nu_\mu + \gamma\,.$$

Another reaction, the radiative pion-β-decay

$$\pi^+ \to \pi^0 e\nu\gamma$$

may come into the range of observation soon, together with new efforts to measure the pion-β-decay constant with better precision as required for a check of the CVC-hypothesis. From the theoretical point of view the radiative pion decay is simpler than $K \to e\nu\pi\gamma$, because to understand the latter, a complete treatment of $SU(3)$ breaking in K_{l3} has still to be provided.

At first sight an additional photon, coupled to a current breaks the symmetry in a way, well to be described by internal symmetries $SU(2)$ or $SU(3)$

$$j_\mu^{\text{el}} = j_\mu^{(V)} + j_\mu^{(S)} = j_\mu^{(3)} + \frac{1}{\sqrt{3}} j_\mu^{(8)} \tag{1.1}$$

where (V) and (S) denote the isovector and isoscalar part, (3) and (8) the corresponding $SU(3)$-indices [3]. Consider however the charged vector current between hadronic states $\langle f | j_\mu^\pm | i \rangle$. It is conserved only as long as there is no photon present besides i and f

$$\langle f | \partial^\mu j_\mu^\pm | i \rangle = O(e^2)$$

so that CVC can be really only true in the subspace of hadronstates and represents thus a *weak* operator equation. On the other hand with an external photon a matrix element

$$\langle f, \gamma | j_\mu^\pm | i \rangle = O(e) \tag{1.2}$$

occurs, but then, because of the "divergence relation" – A_μ is the photon field –

$$\partial^\mu j_\mu^\pm = \mp\, i\, e A^\mu j_\mu^\pm \tag{1.3}$$

the divergence of (1.2) is of the same order e:

$$\langle f, \gamma | \partial^\mu j_\mu^\pm | i \rangle = \pm\, i e \varepsilon_\varrho \langle f | j_\varrho^\pm(0) | i \rangle + O(e^3)\,, \tag{1.4}$$

ε_ϱ is the polarisation vector of the photon. A neutral current is still conserved, thus electromagnetic breaking of the symmetry also implies the appearence of longitudinal components for the matrix elements of the charged currents [4].

After the description of a general method to deal with electromagnetic symmetry breaking in the presence of an external photon in Section 2, we discuss the simplest case, the vector current in $\pi \to e\nu\gamma$, in Section 3. Section 4 is devoted to the more complicated case $\pi^+ \to \pi^0 e\nu\gamma$.

Clearly with the still meager experimental information at hand we must largely rely on models in order to see the main features of the problem.

2. General Considerations

One would be inclined to connect matrix elements of charged and neutral currents by considering appropriate "transverse" parts, as e.g.

$$\langle f\gamma | j_\mu^\pm | i\rangle \mp \frac{ie\,Y_\mu}{(\Delta Y)}\,\varepsilon_\varrho \langle f | j_\varrho^\pm | i\rangle$$

where Δ_μ is the momentum carried by j_μ, and Y_μ any other momentum, and correlate then those matrix elements with the $\langle f\gamma | j_\mu^{(3)} | i\rangle$ by isospin. Apart from the fundamental ambiguity connected with the arbitrary Y, the photon consisting of isoscalar and isovector introduces also gauge conditions, which again only appear for the charged currents at $\Delta^2 \neq 0$. Our equations will be therefore derived from a low energy theorem for any nonchiral internal symmetry, so that – as you heard already in Prof. *Renner*'s lecture – no current algebra is needed. We define

$$T_\lambda^{(k,i)} = i \int \langle f, \gamma(\varepsilon, k) | [j_\lambda^{(k)}(x), O^{(i)}(0)]\, \Theta(x_0) | i\rangle\, e^{-iqx}\, d^4x \qquad (2.1)$$

with any local operator $O^{(i)}(x)$, specified by a certain transformation property under the generators

$$F^{(k)} = \int d^3x\, j_0^{(k)}(x)\,, \qquad \partial_\mu j^{(k)\mu} = 0 \qquad (2.2)$$

of an internal symmetry, which is exact, as long as the electromagnetic interaction is switched off. Further generalizations to chiral symmetries will be obvious. For the charged components (1.3) appears here as

$$\partial^\mu j_\mu^{(k)} = ie\, C^{(k,k')} j_\varrho^{(k')} A^\varrho \qquad (2.3)$$

with appropriate coefficients $C^{(k,k')}$, as soon as the electromagnetic interaction is switched on. Guided by the success of the analogous assumption in current algebra calculations we assume though that

$O^{(i)}$ obeys still

$$[F^k, O^{(i)}] = C_{(j)}^{(k,i)} O^{(j)}. \tag{2.4}$$

Then after a partial integration in $q_\lambda T_\lambda$ in the limit

$$q_\lambda T_\lambda^{(k,i)}|_{q\to 0} = C_{(j)}^{(k,i)} \langle f\gamma | O^{(j)} | i\rangle - U^{(k,i)}, \tag{2.5}$$

$$U^{(k,i)} = -C^{(kk')} i e \varepsilon_\varrho \int e^{ikx} \langle f | [j_\varrho^{(k')}(x), O^{(i)}(0)]\, \Theta(x_0) | i\rangle\, d^4x. \tag{2.6}$$

The limit of the l.h.s. of (2.5) will produce contributions of one particle poles only, if they are degenerate with the mass of one external particle (cf. particle a in Fig. 1).

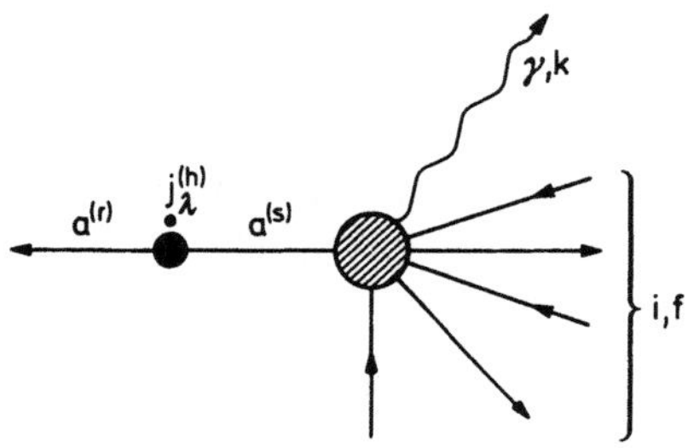

Fig. 1. Pole of particle a, coupled to j_λ

Let such a term be of the type

$$\sum_{\text{Spin},\, a} \frac{q_\lambda \langle a^{(r)}(p') | j_\lambda^{(k)} | a^{(s)}(p'-q)\rangle \langle a^{(s)}(p'-q), f-a | O^{(j)} | i\rangle}{m_a'^2 - (p'-q)^2}.$$

Then, if a has spin zero (e.g. for a pion) and if $C^{(r,k,s)}$ are Clebsch-Gordon-coefficients,

$$\langle a^{(r)} | j_\lambda^{(k)} | a^{(s)}\rangle|_{q\to 0} = 2p'_\lambda C^{(r,k,s)}$$

or for a spinor including the spin summation

$$\sum \langle a^{(r)} | j_\lambda^{(k)} | a^{(s)}\rangle \langle a^{(s)}| = \bar{u}^{(r)} \gamma_\lambda (p' + m)\, C^{(r,k,s)} = 2p'_\lambda \bar{u}\, C^{(r,k,s)}$$

etc. so that with $m_a'^2 - (p'-q)^2 = 2(p'q) - q^2$, for $q \to 0$ a linear equation of the type

$$\begin{aligned} &-C_{(i)}^{(k,j)} \langle f\gamma | O^{(i)} | i\rangle + \sum C^{(r,k,s)} \langle f^{(rs)}\gamma | O^{(j)} | i\rangle \\ &-\sum C^{(r,k,s)} \langle f\gamma | O^{(j)} | i^{(r,s)}\rangle = U^{(k,j)} \end{aligned} \tag{2.7}$$

is obtained. The sums extend over intermediate pole terms in the final and initial state. By means of the superscript (r, s) in these states we have indicated the change of one particle having first the internal index (r) into the index (s). Using all possible commutators and all possible members of the particle multiplets in i and f in this way a system of linear equations is obtained, which usually contains more equations than matrix elements, implying certain relations among the $U^{(k,i)}$. Clearly for $U^{(k,i)}=0$, and the photon state replaced by the vacuum this system must be identical to the symmetry relations for the group applied to $\langle i|j_\lambda^{(k)}|f\rangle$. We note that our technique is identical to the one used in low energy theorems, but instead of expressing the divergence of the axial-vector current through PCAC (ambiguously) by the pion field, the divergence of the vector field gives the purely hadronic matrix element of currents, U, in which the internal symmetry is again exact.

For the use in the vector current of semileptonic radiative weak reactions we pick out that solution of the system (2.7) which determines e.g. the matrix element of the charged operator $O=j_\mu^+$ as a function of the $\langle\gamma f|j_\mu^3|i\rangle$. A certain $U^{(i,k)}$ measures in this case just the difference between matrix elements of charged and neutral conserved currents and an isoscalar photon on one side relative to the full matrix elements containing also nonconserved currents. In a certain sense U may be called therefore the "correction to CVC". These equations contain all symmetry considerations, and the problem of the additional restrictions from gauge invariance etc. can be treated *later* for the specific matrix element to be discussed. We shall see in the next two sections that the U can be sometimes either restricted by internal symmetries or that within a certain *model*, statements on the behaviour of U relative to the matrix elements contained in Eq. (2.7) can be made.

Clearly these relations can be used in other applications also, e.g. to arbitrary "abnormal" currents, like the ones required in certain theorems of CP-violation. In this lecture we restrict ourselves just to the simplest applications in weak interactions, the charged vector currents.

3. Vector Current in $\pi\to e\nu\gamma$

Here $f=0$, $i=\pi^+(p)$, $O\to j_\mu^{(3)}(0)$ and e.g.

$$T_{\lambda\mu}=i\int\langle j(k,\varepsilon)|[j_\lambda^-(x),j_\mu^{(3)}(0)]\,\Theta(x_0)|\pi^+(p)\rangle\,\mathrm{e}^{iqx}\,\mathrm{d}^4x\,. \tag{3.1}$$

From the crossed term only the π^0 intermediate state ($m_{\pi^0}=m_{\pi^+}$) contributes to $q_\lambda T_{\lambda_\mu}|_{q\to 0}$, so that with

$$[I^-,j_\mu^3(0)]=j_\mu^-(0)\,,$$
$$\langle\pi^0(p)|j_\lambda^-(0)|\pi^+(p)\rangle=\sqrt{2}\,2p_\lambda\,,$$

and Eq. (1.3) simply

$$\begin{aligned}-\sqrt{2}\langle\gamma|j_\mu^3|\pi^0\rangle&=\langle\gamma|j_\mu^-|\pi^+\rangle-U_\mu\,,\\ U_\mu&=ie\varepsilon_\varrho\int\mathrm{e}^{-ikx}\langle 0|[j_\varrho^-(x),j_\mu^{(3)}(0)]\,\Theta(x_0)|\pi^+(p)\rangle\,\mathrm{d}^4x\end{aligned} \tag{3.2}$$

follows. Let $j_\mu^{(3)}$ in U_μ and of course also j_μ^3 and j_μ^- in the first matrix elements of (3.2) too – be a current of first kind, as j_ϱ^- in U must be anyhow, then $U \equiv 0$ from G-parity and "CVC" holds exactly. Actually this result could have been obtained much more easily by the observation that by the same G-parity argument the photon in $\langle\gamma|j|\pi\rangle$ must be a pure isoscalar so that again the relative factor is just a Clebsch-Gordon-coefficient. On the other hand longitudinal components of the current are here absent anyhow because parity conservation makes $\langle\gamma|\partial j|\pi\rangle = \langle\gamma|j|\pi\rangle_{\varepsilon=k} = 0$. Because of the symmetry of the photons in the amplitude for $\pi^0 \to 2\gamma$

$$\begin{aligned}\langle\gamma|j_\mu^{\mathrm{el}}|\pi^0\rangle &= \langle\gamma|j_\mu^{(\mathrm{S})} + j_\mu^{(\mathrm{V})}|\pi^0\rangle \\ &= \langle\gamma^{(\mathrm{V})}|j^{(\mathrm{S})}|\pi^0\rangle + \langle\gamma^{(\mathrm{S})}|j_\mu^{(\mathrm{V})}|\pi^0\rangle = 2\langle\gamma|j_\mu^{(3)}|\pi^0\rangle\end{aligned} \tag{3.3}$$

the left hand of Eq. (3.2) is expressible in terms of the π^0-decay rate. The matrix element $\langle\gamma|j_\mu^-|\pi^+\rangle$ can be extracted from the $\pi \to e\nu\gamma$-rate [5] if the corresponding axialvector current is determined independently either by an appropriate correlation measurement or from theoretical arguments [6]. A deviation from (3.2) with $U = 0$ would thus indicate an admixture of currents of 2nd kind in j^μ.

4. The Vector Current in $\pi^+ \to \pi^0 e\nu\gamma$

We treat this reaction here mainly because of its simplicity. Even after observing this reaction, as long as Compton scattering of pions cannot be observed separately, the relations to be derived are of use to clarify the "CVC-deviation" referred to in the introduction within a specific model.

The vector current in radiative μ-capture

$$\mu^- + p \to n + \nu_\mu + \gamma$$

is really a much more realistic example, but also there the experimental information is still rather limited [2] and the complications due to spin decrease the pedagogical value for a presentation in this lecture [9].

To illustrate our remarks in Section 2 we give first the general system of linear relations between all matrix elements $\langle\gamma\pi|j|\pi\rangle$, Eq. (2.7). It is obtained by writing down for $T_{\lambda\mu}$ all possible retarded commutators of $SU(2)$ currrnts, with q attached to either current and sandwiched between all states $\langle\pi\gamma||\pi\rangle$ which are allowed by charge conservation. Clearly those relations are trivial where $j_\lambda^3(x)$ appears. In terms of the abbreviations

$$\begin{aligned}\langle\gamma\pi|j^{(i)}|\pi^{(k)}\rangle &= (i,k)\,, \qquad i,k = 0,+,-\,, \\ -\int\langle\gamma\pi|[\partial j^{(i)}(x), j_\mu^{(j)}(0)]\,O(x_0)|\pi^{(k)}(p)\rangle\,\mathrm{d}^4x &= (i,j,k)\end{aligned} \tag{4.1}$$

the nontrivial relations are

$$
\begin{aligned}
(-,+) &= \sqrt{2}(0,+) - \sqrt{2}\,(0,0) + (-,0,+)\\
(-,0) &= \sqrt{2}(0,0) - \sqrt{2}\,(0,-) + (-,0,0)\\
(+,0) &= \sqrt{2}(0,-) - \sqrt{2}\,(0,0) + (+,0,-)\\
(+,0) &= \sqrt{2}(0,0) - \sqrt{2}\,(0,+) + (+,0,0)\\
-\sqrt{2}(0,0) &= (+,0) + (+,-) + \frac{1}{\sqrt{2}}(-,+,0)\\
\sqrt{2}(0,0) &= (-,0) - (-,+) + \frac{1}{\sqrt{2}}(+,-,0)\\
\sqrt{2}(0,+) &= -(+,0) + \frac{1}{\sqrt{2}}(-,+,+)\\
\sqrt{2}(0,+) &= (-,+) + \frac{1}{\sqrt{2}}(+,-,+)\\
-\sqrt{2}(0,-) &= (+,-) + \frac{1}{\sqrt{2}}(-,+,-)\\
\sqrt{2}(0,-) &= -(-,0) + \frac{1}{\sqrt{2}}(+,-,-)
\end{aligned}
\tag{4.2}
$$

including 10 equations for the three isovector-Compton amplitudes $(0,+)$, $(0,-)$, $(0,0)$ and for the 4 charged currents $(-,+)$, $(+,-)$, $(-,0)$, $(+,0)$. The first four equations also hold with the $j^{(3)}$ everywhere replaced by j^{el}. They relate the $\langle|j^{\pm}|\rangle$ to the full Compton amplitudes. It is easily seen that four identities hold among the U's, so that all current matrix elements can be expressed in terms of the U's and one current matrix element, e.g. $(-,0)$, $(0,0)$ being a combination of U only. This is consistent with the fact that for $U=0\,((\partial j^{\pm}=0)$ and $SU(2))$ all matrix elements can be expressed in terms of one by Clebsch-Gordon-coefficients only, except for $(0,0)$, which must become zero. We do not bother to go further into details here, because the relations (4.2) will not be used below with the exception of e.g. the eight relation

$$M_\mu^- = M_\mu^0 + U_\mu \tag{4.3}$$

where

$$
\begin{aligned}
M_\mu^- &= \langle j(k,\varepsilon_\varrho), \pi^0(p')|j_\mu^-|\pi^+(p)\rangle\,,\\
M_\mu^0 &= \langle \gamma\pi|j_\mu^3|\pi^+\rangle\,,
\end{aligned}
\tag{4.4}
$$

$$U_\mu = e\varepsilon_\varrho U_{\varrho\mu} = \frac{ie\varepsilon_\varrho}{\sqrt{2}} \int e^{ikx} \langle\pi^+(p')|[j_\varrho^+(x), j^-(0)]\,\Theta(x_0)|\pi^+(p')\rangle\, d^4x\,.$$

We note that M^- comes here from a $\pi^0\gamma$ intermediate state in $q_\lambda T_{\lambda\mu}$, where the photon is disconnected.

In general a matrix element like M^- must fulfill a Ward-Takahashi identity [10] because it is not gauge invariant by itself. This identity is most simply derived in a model, e.g. the "realistic" one with a leptonic current attached to j_μ^-. Then the diagram a) of Fig. 2 is gauge invariant

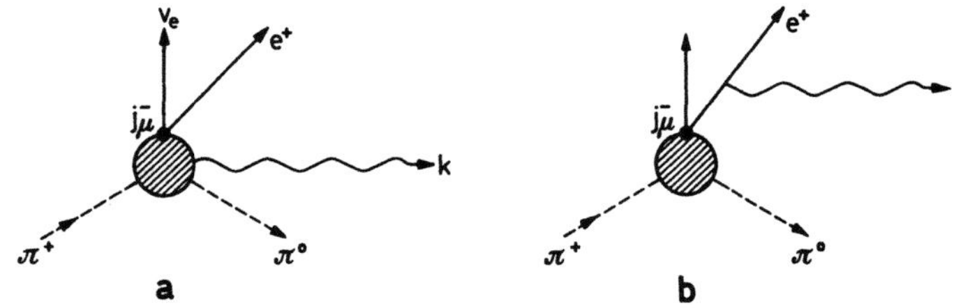

Fig. 2. The two diagrams $eM_\mu^- j_\mu$ and $M^{\text{in.b.}}$ for the vector current in $\pi \to \pi e\nu\gamma$

only together with the internal bremsstrahlung, diagram b). Omitting the leptonic current, which becomes a common factor, this yields

$$M_\mu^- |_{\varepsilon=k} = -\langle \pi^0 | j_\mu^- | \pi^+ \rangle . \tag{4.5}$$

An obvious separation of M^{total} into separately gauge invariant contributions would be

$$M^{\text{total}} = ej^\mu \left[M_\mu^- + \frac{(\varepsilon x)}{(kx)} \langle \pi^0 | j_\mu^- | \pi^+ \rangle \right] + M^{\text{in.b.}} - \frac{(\varepsilon x)}{(kx)} \langle \pi^0 | j_\mu^- | \pi^+ \rangle \, ej^\mu . \tag{4.6}$$

Here X can be any linear combination of p and p'. For the axial vector part because of parity no contribution comes from $M^{\text{in.b.}}$ and M_μ^{axial} is gauge invariant by itself. There also from G-parity $U^{\text{axial}} \equiv 0$. The isovector part of pion isovector-Compton-scattering $M^0/\sqrt{2}$ is gauge invariant by itself, thus from (4.3) and (4.5a)

$$k_\varrho U_{\varrho\mu} = -\langle \pi^0 | j_\mu^- | \pi^+ \rangle \tag{4.7}$$

is obtained. Note that this equation is derived only from gauge invariance and the divergence Eq. (1.3). Current algebra

$$[j_0^+(x), j_\mu^-(0)]\, \delta(x_0) = 2 j_\mu^3 \, \delta^4(x)$$

together with

$$\sqrt{2} \langle \pi^+ | j^{(3)} | \pi^+ \rangle = \langle \pi^0 | j^- | \pi^+ \rangle$$

is *sufficient* to prove it also. Any sum rule, based on (4.7) is therefore not at all a test of current algebra. From its derivation as a gauge condition we rather think that (4.7) *determines* "subtraction" terms (cf. below).

A second restriction comes from the prescribed way, longitudinal components appear in j_μ. From Eqs. (1.3) we derive ($\Delta = p - p' - k$)

$$\Delta_\mu M_\mu^- = ie\langle \gamma\pi^0 | \partial j^-(0) | \pi^+ \rangle = e\varepsilon_\varrho \langle \pi^0 | j_\varrho^-(0) | \pi^+ \rangle \tag{4.5a}$$

and hence with (4.3)

$$\Delta_\mu U_{\varrho\mu} = \langle \pi^0 | j_\varrho^-(0) | \pi^+ \rangle . \tag{4.7a}$$

Before considering a simple model for radiative $\pi - \beta$-decay, we indicate the corresponding relation in the (already observed) decay $K^+ \to e\nu\pi^0\gamma$. Starting from

$$T_{\lambda\mu}^K = i \int \langle \pi^+ \gamma | [j_\lambda^+(x), (j_\mu^{(4)} - ij_\mu^{(5)})] \, \Theta(x_0) | K^+ \rangle \, e^{-iqx} \, d^4x$$

we would arrive at

$$\sqrt{2}\langle \pi^0\gamma | (j_\mu^{(4)} - ij_\mu^{(5)}) | K^+ \rangle = -\langle \pi^+ \gamma | (j_\mu^{(6)} - ij_\mu^{(7)}) | K^+ \rangle$$
$$+ ie\varepsilon_\varrho \int \langle \pi^+ | [j_\varrho^+(x), (j_\mu^{(4)} - ij_\mu^{(5)})] \, \Theta(x_0) | K^+ \rangle \, e^{ikx} \, d^4x .$$

Only by the assumption of exact $SU(3)$ the neutral, but strangeness carrying, current can be related to a pion Compton scattering as before

$$\langle \pi^+ \gamma | (j^{(6)} - ij^{(7)}) | K^+ \rangle = -\langle \pi^+ \gamma | j^{(3)} | \pi^+ \rangle + \langle \pi^+ \gamma | j^{(5)} | \pi^+ \rangle .$$

Only a better understanding of the actual $SU(3)$ breaking in the $K\pi$-system would lead to a more realistic relation.

In order to obtain a better feeling for the "deviation" U relative to M^0 in Eq. (4.3) despite the absence of experimentally measured quantities, which would be fed more directly in M^- and M^0 of this relation, we now turn to a simple dispersion theoretical model in U and M^0, retaining only the lowest single particle states.

For this purpose we develop as usual ($\Delta = p - p' - k$, $P = (p + p')/2$)

$$\begin{aligned} M_\mu^0 &= e\varepsilon_\varrho M_{\varrho\mu} , \\ M_{\varrho\mu} &= R_0 g_{\varrho\mu} + P_\varrho [P_\mu R_1 + k_\mu R_2 + \Delta_\mu R_3] \\ &\qquad \Delta_\varrho [P_\mu R_1' + k_\mu R_2' + \Delta_\mu R_3'] \end{aligned} \tag{4.8}$$

and we use a similar decomposition in terms of U_0, U_i, U_i' for $U_{\varrho\mu}$, Eq. (4.5). In the case of the latter the absorptive part reads

$$\begin{aligned} u_{\varrho\mu} \equiv \text{abs}(U_{\varrho\mu}) = \frac{(2\pi)^4}{2\sqrt{2}} \Big[&\sum_n \langle \pi^+ | j_\varrho^+ | n \rangle \langle n | j_\mu^- | \pi^+ \rangle \, \delta^4(p' + k - P_n) \\ &- \sum_m \langle \pi^+ | j_\mu^- | m \rangle \langle m | j_\varrho^+ | \pi^+ \rangle \, \delta^4(p - k - P_m) \Big] . \end{aligned} \tag{4.9}$$

For simplicity we disperse as in Ref. [9] in terms of

$$\nu = (pk) = (p'k) \tag{4.10}$$

so that

$$\begin{aligned} &\Delta^2=(p-p'-k)=t=(p-p')^2\,,\\ &(\Delta k)=0\,. \end{aligned} \tag{4.11}$$

The contribution of the π^0-state in (4.9)

$$U^{(\pi)}_{\varrho\mu}=-\frac{\sqrt{2}F_\pi(t)}{\nu}P'_\varrho(p_\mu+k_\mu+p'_\mu) \tag{4.12}$$

plays a special role in the gauge condition (4.7)

$$k_\varrho U_{\varrho\mu}=-2\sqrt{2}F_\pi(t)\,P_\mu$$

and in (4.7a)

$$\Delta_\mu U_{\varrho\mu}=2\sqrt{2}F_\pi(t)\,P_\varrho$$

with $U=U^\pi+\tilde{U}$ the two conditions

$$\begin{aligned} k_\varrho\tilde{U}_{\varrho\mu}&=\sqrt{2}F_\pi(t)\,k_\mu\,,\\ \Delta_\mu U_{\varrho\mu}&=\sqrt{2}F_\pi(t)\,\Delta_\varrho \end{aligned} \tag{4.13}$$

follow. We consider furthermore the intermediate ω and A_1-state only

$$\begin{aligned} \langle\pi^0(p')|j^{(3)}_\mu|\omega(Q)\rangle&=-\frac{1}{\sqrt{2}}\langle\pi^+(p')|j^+_\mu|\omega(Q)\rangle\\ &=f_{\omega\pi\gamma}\varepsilon_{\mu\alpha\beta\gamma}P'_\alpha Q_\beta\varepsilon^{(\omega)}_\gamma\,,\\ \langle\pi^+(p')|j^{(3)}_\mu|A^+(Q)\rangle&=\frac{1}{\sqrt{2}}\langle\pi^+|j^+_\mu|A^0\rangle\\ &=f_{A\pi\gamma}\left[g_{\lambda\mu}+\frac{P'_\lambda(P'_\mu+Q_\mu)}{m_A^2-m_\pi^2}\right]\varepsilon^{(A)}_\lambda \end{aligned} \tag{4.14}$$

yielding contributions

$$\begin{aligned} U^{(\omega)}_{\varrho\mu}=\frac{\sqrt{2}|f_{\omega\pi\gamma}|^2}{m_\omega^2-m_\pi^2-2\nu}&\left\{\frac{m_\omega^2-m_\pi^2}{4}g_{\varrho\mu}[m_\omega^2-m_\pi^2+t]\right.\\ &+P'_\varrho\left[p_\mu\frac{m_\omega^2-m_\pi^2}{2}-k_\mu\frac{m_\omega^2+m_\pi^2}{2}\right]\\ &\left.-P'_\varrho\left[p'_\mu\frac{m_\omega^2-m_\pi^2}{2}-k_\mu\left(m_\pi^2-\frac{t}{2}\right)\right]\right\}, \end{aligned} \tag{4.15}$$

$$\begin{aligned} U^{(A)}_{\varrho\mu}=-\frac{\sqrt{2}|f_{A\pi\gamma}|^2}{m_A^2-m_\pi^2-2\nu}&\left\{g_{\varrho\mu}+\frac{P_\varrho}{m_A^2-m_\pi^2}(P_\mu+k_\mu+p'_\mu)\right.\\ &\left.-\frac{P'_\varrho}{m_A^2-m_\pi^2}(P_\mu+3k_\mu+p'_\mu)\right\}. \end{aligned} \tag{4.16}$$

We shall be modest enough not to assume complete saturation of (4.13) by (4.15) and (4.16). We rather introduce the smallest order polynomials in the smallest number of invariants, just sufficient to be determined by (4.13) representing "seagull" terms or "subtraction constants". These turns out to be two additional constants U_{0C} in U and U'_{3C} in U_3 besides the other pole contributions:

$$\begin{aligned}\sqrt{2}F_\pi(t) &= U_{0C} + \sqrt{2}|f_{\omega\pi\gamma}|^2 \frac{m_\omega^2 - m_\pi^2 + t}{4} - \sqrt{2}\frac{|f_{A\pi\gamma}|^2}{m_A^2 - m_\pi^2} + \cdots \\ 0 &= U'_{3C} - \sqrt{2}\frac{|f_{\omega\pi\gamma}|^2}{4} + \cdots .\end{aligned} \tag{4.17}$$

U_{0C} and U'_{3C} can be considered to represent in a certain sense the respective continuum integrals, apart from the ω- and A_1-pole. With the *assumption* $U_{0C} \equiv 0$, the first relation (4.17) leads to the sum rule of Ref. [11]. For us it will be enough that e.g. U_{0C} is determined by (4.17) so that $(m_\pi \ll m_{\omega_1} m_A)$

$$U_0 = \sqrt{2}F_\pi + \frac{\sqrt{2}}{2}|f_{\omega\pi\gamma}|^2 m_\omega^2 \frac{\nu}{m_\omega^2 - 2\nu} - 2\sqrt{2}|f_{A\pi\gamma}|^2 \frac{\nu}{m_A^2(m_A^2 - 2\nu)}. \tag{4.18}$$

The other U_i, U'_i can be read off easily from (4.12), (4.15), (4.16). Our considerations are completely independent from the size of U_{0C}. Therefore the (doubtful) "sum rule assumption" $U_{0C} = 0$ would lead to the same U_0, Eq. (4.18).

In the same way the Compton amplitude M^0 can be calculated, here (4.7) and (4.7a) must be replaced by

$$\begin{aligned} k_\varrho M_{\varrho\mu} &= 0, \\ \Delta_\mu M_{\varrho\mu} &= 0, \end{aligned} \tag{4.19}$$

and because of the two neutral currents in $\mathrm{abs}(M_{\varrho\mu})$ the ω does not contribute. With the minimal number of "subtractions" as required by (4.19) we need only a constant R_{0C} in R_0 yielding

$$R_0/\sqrt{2} = -2F_\pi(t) - 2\frac{|f_{A\pi\gamma}|^2}{m_A^2}\frac{4\nu^2}{m_A^2 - 4\nu^2}. \tag{4.20}$$

The other invariants may be picked out easily from the contributions

$$\begin{aligned} M_{\varrho\mu}^{(\pi)} &= \frac{\sqrt{2}F_\pi}{\nu}[p_\varrho(p_\mu - k_\mu + p'_\mu) - p'_\varrho(p_\mu + k_\mu + p'_\mu)] \\ M_{\varrho\mu}^{(A)} &= -\frac{\sqrt{2}|f_{A\pi\gamma}|^2}{m_A^2 - m_\pi^2}\left[\frac{p_\varrho(p_\mu + k_\mu + p'_\mu) - p'_\varrho(p_\mu + 3k_\mu + p'_\mu) + (m_A^2 - m_\omega^2)}{m_A^2 - m_\pi^2 - 2\nu} g_{\varrho\mu}\right. \\ &\quad \left. + \frac{p_\varrho(-p_\mu + 3k_\mu - p'_\mu) + p'_\varrho(p_\mu - k_\mu + p'_\mu) + (m_A^2 - m_\omega^2)}{m_A^2 - m_\pi^2 - 2\nu} g_{\varrho\mu}\right]. \end{aligned} \tag{4.21}$$

The kinematical range of the reaction $\pi^+ \to \pi^0 e\nu\gamma$ determines the relative importance of M^0 and U. Because of the small energy release $Q = m_{\pi^+} - m_{\pi^0} \cong 4.6$ MeV, $0 \leqq k_0 \leqq O(Q)$, we have besides $p = (m_\pi, \mathbf{O})$ also $p' = (m_\pi, \mathbf{O}) + O(Q)$. Inspection of Eqs. (4.15) and (4.21) shows that therefore all terms contribute only to $O\left(\frac{Q m_\pi}{m_A^2}\right)$ or $O\left(\frac{Q m_\pi}{m_\omega^2}\right)$, *except* for U_0 and R_0. Furthermore because $0 \leqq \nu \leqq m_\pi Q \sim (20\text{ MeV})^2$ values of $r \ll m_A^2, m_\omega^2, m_\pi^2$ occur only. Near $\nu = 0$ the pion contribution dominates both $U_0|_{\nu=0} = \sqrt{2}$ and $R_0|_{\nu=0} = -2\sqrt{2}$. The corrections can be easily estimated: $f_{\omega\pi\gamma}$ is determined by $\Gamma_{\omega\to\pi\gamma}$; $f_{A\pi\gamma}$ can be related to the longitudinal parameter in $A \to \varrho\pi$ decay by a hard pion technique [13] and a value for the latter exists now in the literature [14]. As expected from the shape of (4.18) and (4.20) the corrections are in fact of the order of 1% for the maximal value of ν, much below the accuracy of the model anyhow. We find therefore that the "correction" U_0 is of the same order of magnitude as the "Compton" matrix element R_0. It is completely determined at $\nu = 0$ by the "seagull" term of the intermediate one-pion-state. This coincides with the known results of low energy limits [15], in our approach however also the corrections for $\nu \neq 0$ can be determined.

Acknowledgements

The author acknowledges gratefully the help of *W. Majerotto* and *M. Schweda.*

References and Footnotes

1. *Barash-Schmidt, N., Barbaro-Galtieri, A., Conforto, G., Price, L. R., Roos, M., Rosenfeld, A. H., Söding, P., Wohl, C. G.:* Particle data group, edition January 1969. Berkeley, L.R.L.
2. *Conforto, G., Conversi, M., di Lella, L.:* Phys. Rev. Letters **9**, 22 (1962).
   Conversi, M., Diebold, R., di Lella, L.: CERN 1963 (unpublished).
3. We use the metric $g_{00} = 1, g_{ij} = -\delta_{ij}, (i,j = 1, 2, 3)$, our states are normalized as $\langle \mathbf{k}' | \mathbf{k} \rangle = 2k_0 (2\pi)^3 \delta^3(\mathbf{k} - \mathbf{k}')$ and spinors as $\bar{u}u = 2m$.
4. *Kummer, W., Majerotto, W.:* Nuovo Cimento A **55**, 558 (1968).
5. *Depommier, P., Heintze, J., Rubbia, C., Soergel, V.:* Phys. Letters **7**, 285 (1963).
6. Note that for the axialvector there are – apart from the internal bremsstrahlung – predictions from the quark model [7] (that it should vanish) or from hard pion techniques [8] in terms of decay parameters of the A_1-meson.
7. *Kummer, W., Kuti, J.:* Nuovo Cimento A **45**, 242 (1966).
8. e.g. *Brown, S. C., West, G. B.:* Phys. Rev. **168**, 1605 (1968).
9. *Kummer, W., Schweda, M.:* to be published.
10. *Takahashi, Y.:* Nuovo Cimento **6**, 370 (1957).
11. *Fayyazuddin, Riazuddin:* Phys. Rev. Letters **18**, 715 (1967). From Regge pole behaviour at high ν, as used in superconvergence considerations one would in contradiction to this ref. assume that U_0 requires a subtraction. Also the proof of equivalence of a corresponding sume rule with the Harari-Pagels-sum rule [12] seems to us not quite correct.
12. *Pagels, H.:* Phys. Rev. Letters **18**, 316 (1967). — *Harari, H.:* Phys. Rev. Letters **18**, 319 (1967).

13. *Konetschny, W., Kummer, W.*: Nuovo Cimento (to be published).
14. *Brown, S. C., West, G. B.*: Quote in a recent preprint a correction to the value of: *Ballam, J., Brody, A. D., Chadwick, G. B., Fries, D., Guiragessián, Z. G. T., Johnson, W. B., Larsen, R. R., Leith, D. W. G. S., Martin, F., Perl, M., Pickup, E., Tan, T. H.*: Phys. Rev. Letters **21**, 934 (1968), namely for the ratio of the longitudinal and transversal rate $(g_L/g_T) = 0.64$.
15. *Adler, S. L., Dothan, Y.*: Phys. Rev. **151**, 1267 (1966).

Dr. *W. Kummer*
Institut für Hochenergiephysik der Österreichischen Akademie der Wissenschaften
A-1050 Wien, Nikolsdorfergasse 18
and
Institut für Theoretische Physik der Technischen Hochschule Wien

Radiative Corrections to Weak Decays Involving Leptons

RIAZUDDIN

Contents

1. Introduction 126
2. Radiative Corrections to the β-Decay of Point-like Particles 129
3. Radiative Corrections to $\Delta Y=0$ Semi-leptonic Processes 136
a) Divergence Problems 136
b) Elimination of Divergences in Certain Models 145
c) Numerical Results 148
d) Applications. 150
4. Radiative Corrections to $|\Delta Y|=1$ Semi-leptonic Processes 156
a) Divergences in Radiative Corrections to $|\Delta Y|=1$ Processes 156
b) Radiative Corrections to K_{l3} Decay. 157
c) Radiative Corrections to $K_{e2}^{\pm}/K_{\mu 2}^{\pm}$ Ratio 158
References and Footnotes. 159

1. Introduction

In these lectures I shall discuss the problem of radiative corrections to leptonic weak decays. We shall be concerned with the computing of *second order* radiative corrections to *first order* weak interactions. We shall regard weak interactions to be local interactions of two currents, i.e. we write

$$\mathscr{L}_{\mathrm{W}}^{\mathrm{l.}}=\frac{G}{\sqrt{2}}L_\lambda \bar{L}_\lambda ,$$

$$\mathscr{L}_{\mathrm{W}}^{\mathrm{s.l.}}(\Delta Y=0)=\frac{G^0}{\sqrt{2}}(J_\lambda^0\bar{L}_\lambda+\bar{J}_\lambda^0 L_\lambda), \qquad (1)$$

$$\mathscr{L}_{\mathrm{W}}^{\mathrm{s.l.}}(|\Delta Y|=1)=\frac{G^1}{\sqrt{2}}(J_\lambda^1\bar{L}_\lambda+\bar{J}_\lambda^1 L_\lambda),$$

where

$$L_\lambda=i\bar{\nu}_e\gamma_\lambda(1+\gamma_5)e+i\bar{\nu}_\mu\gamma_\lambda(1+\gamma_5)\mu ,$$
$$\bar{L}_\lambda=L_\lambda^+(1-2\delta_{\lambda 4})$$

and J_λ^0 and J_λ^1 are, respectively, $\Delta Y=0$ and $\Delta Y=1$ hadronic currents. The following properties of J_λ^0 and J_λ^1 are fairly well established:

(i) $$J_\lambda^0 = V_\lambda^0 + A_\lambda^0, \qquad J_\lambda^1 = V_\lambda^1 + A_\lambda^1,$$

(ii) V_λ^0, $\bar{V}_\lambda^0$, $j_\lambda^{\text{e.m.}}$ $(I=1)$ are $1+i2$, $1-i2$ and 3 members of isospin current $\mathscr{I}_{i\lambda}(i=1,2,3)=V_{i\lambda}$ which is conserved by strong interactions (CVC). The charge operators

$$I_\pm(t) = -i\int V_{1\pm i2,4}(\mathbf{x},t)\mathrm{d}^3x\,,$$
$$I_3(t) = -i\int V_{3,4}(\mathbf{x},t)\mathrm{d}^3x$$

form algebra of $SU(2)$:

$$[I_i, I_j] = i\varepsilon_{ijk}I_k$$

(iii) The $\Delta Y=0$ charges $I_\pm$, I_3 and

$$I_\pm^5(t) = -i\int A_{1\pm i2,4}(\mathbf{x},t)\mathrm{d}^3x$$
$$I_3^5(t) = -i\int A_{3,4}(\mathbf{x},t)\mathrm{d}^3x$$

generate algebra of $SU(2)\otimes SU(2)$.

(iv) V_λ^0 and V_λ^1 are the $1+i2$ and $4+i5$ members of an octet of vector currents $V_{i\lambda}$ $(i=1,\ldots,8)$ and A_λ^0 and A_λ^1 are the $1+i2$ and $4+i5$ members of an octet of axial vector currents $A_{i\lambda}$. Their commutation relations at equal time are assumed to be [1]

$$\delta(x_0)[V_{i0}(x), V_{j\lambda}(0)] = \delta(x_0)[A_{i0}(x), A_{j\lambda}(0)] = if_{ijk}V_{k\lambda}(0)\delta^4(x)\,, \tag{2a}$$

$$\delta(x_0)\,[V_{i0}(x), A_{j\lambda}(0)] = \delta(x_0)[A_{i0}(x), V_{j\lambda}(0)] = if_{ijk}A_{k\lambda}(0)\delta^4(x)\,. \tag{2b}$$

The commutation relations (2a) for $\lambda \neq 4$ contain c-number Schwinger terms [2]. If, however, one integrates V_{i0} and A_{i0} over d^3x, the Schwinger terms are integrated out. In other words, the commutation relations (2) imply that $\Delta Y=0$ and $\Delta Y=\pm 1$ charges generate algebra of $SU(3)\otimes SU(3)$. Many tests of the algebraic structure (2) have been successfully performed [1], especially in conjunction with the PCAC hypothesis like Adler-Weisberger relation and a large class of soft pion theorems.

v) Cabibbo form of universality of weak interactions

$$G^0 = G\cos\theta, \qquad G^1 = G\sin\theta\,. \tag{3}$$

With this form of universality, using (1), we can write for weak processes involving leptons the following phenomenological Lagrangian:

$$\mathscr{L}_{\mathrm{W}} = \frac{G}{\sqrt{2}} [L_\lambda \bar{L}_\lambda + J_\lambda \bar{L}_\lambda + \bar{J}_\lambda L_\lambda] \tag{4a}$$

where

$$\begin{aligned} J_\lambda &= \cos\theta J_\lambda^0 + \sin\theta J_\lambda^1 \\ &= \cos\theta J_{1+i2,\lambda} + \sin\theta J_{4+i5,\lambda}, \\ \bar{J}_\lambda &= \cos\theta J_{1-i2,\lambda} + \sin\theta J_{4-i5,\lambda}. \end{aligned} \tag{4b}$$

The computation of radiative corrections to μ-decay and semi-leptonic processes is necessary for the comparison of theoretical predictions with experiment. One instance where corrections of a few per cent are very important is the question of universality. From the observed μ-life-time (from which the radiative corrections have not been subtracted out) one finds the weak coupling constant to be [3]

$$G^\mu = (1.4320 \pm 0.0011) \times 10^{-49} \text{ erg cm}^3 \tag{5}$$

while from observed O^{14} decay rate (which is corrected for *Coulomb* [4] corrections but from which radiative corrections have not been subtracted out) one finds the weak coupling constant to be [4]

$$G_V^\beta = (1.4173 \pm 0\,0034) \times 10^{-49} \text{ erg cm}^3 . \tag{6}$$

These coupling constants agree within a few per cent. Now CVC implies that G^0 is not renormalized by strong interactions. Since CVC is violated by electromagnetic interactions there will be electromagnetic corrections to G^0 (G also suffers similar corrections). Thus if one assumes the bare coupling constants (i.e. the coupling constants in the absence of electromagnetism) in μ- and β-decays to be equal, i.e. $G = G^0$, then CVC implies that any difference between G^μ and G_V^β must be due to charge dependent effects. In the Cabibbo form of universality G^0 is not equal to G but to $G\cos\theta$. Therefore, if after subtracting out the radiative corrections from the observed quantities G^μ and G_V^β, G and G^0 still differ, then such a difference can be attributed to $\cos\theta$. Another nice thing about the Cabibbo form of universality is that it suppresses $|\Delta Y| = 1$ decays with respect to $\Delta Y = 0$ decays as seen experimentally. Thus θ as determined from $K^+ \to \pi^0 e^+ \nu_e$ decay rate is given by [5]

$$\theta \approx \sin\theta = (0.230 \pm 0.003) . \tag{7}$$

This would imply that G^0 differs from G by about 2%, which is of the same order as one would expect from second-order radiative corrections.

Therefore, in order to establish that $G^0 = G\cos\theta$, one must have a reliable estimate of the radiative corrections.

Another point about these corrections is whether they are finite or infinite. As far as μ-decay is concerned, they are finite as we shall see. For semi-leptonic decays (e.g. $n \to p + e^- + \bar{\nu}_e$), they are logarithmically divergent if n and p are treated as point-like particles. The question then arises whether the structure of hadrons involved could lead to convergent results. We shall explore this possibility in detail. For this purpose we shall use the Bjorken technique [6], which expresses the asymptotic behaviour of an amplitude of the form

$$T_{\mu\lambda} = i \int d^4x\, e^{-iQ\cdot x} \langle B(p')| T(j_\mu(x) J_\lambda(0)) |A(p)\rangle \tag{8}$$

which will be encountered in the radiative corrections to β-decay of a hadron when strong interactions are considered to all orders, as follows:

$$\begin{aligned} T_{\mu\lambda} \underset{\substack{Q_0\to\infty \\ \boldsymbol{Q}=0}}{\longrightarrow} & -\frac{1}{Q_0} \int d^3x \langle B| [j_\mu(\boldsymbol{x},0), J_\lambda(0)] |A\rangle \\ & -\frac{1}{Q_0^2} \int d^3x \langle B| [[j_\mu(\boldsymbol{x},0), H], J_\lambda(0)] |A\rangle \,. \end{aligned} \tag{9}$$

Thus it will turn out that the question of divergence or convergence of the radiative corrections (at least to vector coupling constant) does not depend on the details of strong interaction but rather on the structure of the current algebra. Simple realizations of such an algebra are usually obtained in terms of free fields, corresponding to point-like particles. The computation of radiative corrections for point-like particles will therefore provide us with an insight into the final outcome, convergent or divergent, of computing radiative corrections to the β-decays of real hadrons.

2. Radiative Corrections to the β-Decay of Point-like Particles

First we discuss the radiative corrections to μ-decays and, more generally, to the case of the β-decay of a point-like particle of arbitrary charge.

The radiative corrections to μ-decay were calculated by *Berman* [7] and by *Kinoshita* and *Sirlin* [8]. They turn out to have no ultraviolet divergence in the V-A theory. This can be seen as follows. The reduced T-matrix element ($S = 1 + iT$) for the μ-decay

$$\mu^-(p) \to \bar{e}(p') + \nu_\mu(k_1) + \bar{\nu}_e(k_2) \tag{10}$$

is given by

$$T_0 = -\frac{G}{\sqrt{2}}[\bar{u}(k_1)\gamma_\lambda(1+\gamma_5)u(p)][\bar{u}(p')\gamma_\lambda(1+\gamma_5)v(k_2)]\cdot N \qquad (11)$$

where N is the normalization factor $(2\pi)^{-6}\sqrt{m_\mu m_e m_{\nu_\mu} m_{\nu_e}/p_0 p'_0 k_{10} k_{20}}$. This corresponds to the interaction of the form (V-A)

$$\mathscr{L}_{\mathrm{W}}^{\mu}(x) = \frac{G}{\sqrt{2}}[\bar{\nu}_\mu(x) i\gamma_\lambda(1+\gamma_5)\mu(x)][\bar{e}(x) i\gamma_\lambda(1+\gamma_5)\nu_e(x)] + \text{h.c.} \qquad (12)$$

In lowest order, the Feynman diagrams corresponding to the radiative corrections to μ-decay are shown in Fig. 1.

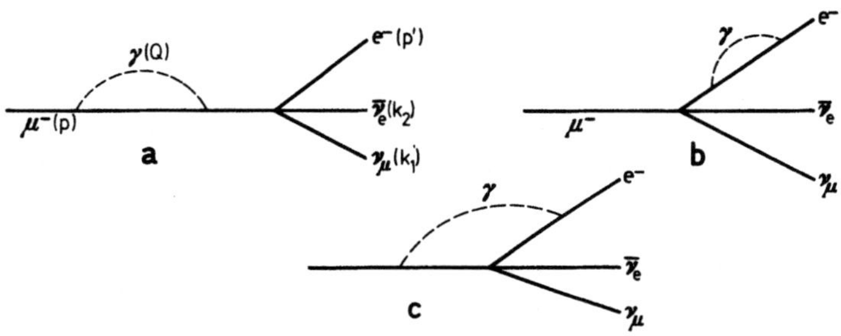

Fig. 1. Feynman diagrams for lowest order radiative corrections to μ-decay

The correction arising from the diagrams a, b and c is given by

$$T_1 = T_{1a} + T_{1b} + T_{1c} \qquad (13)$$

with

$$T_{1a} = -\frac{G}{\sqrt{2}}[\bar{u}(p')\gamma_\lambda(1+\gamma_5)v(k_2)] \times \bar{u}(k_1)\gamma_\lambda(1+\gamma_5)\frac{1}{i\gamma\cdot p + m_\mu}\Sigma^{(2)}(p)u(p)\cdot N\,, \qquad (14a)$$

$$T_{1b} = -\frac{G}{\sqrt{2}}\bar{u}(p')\Sigma^{(2)}(p')\frac{1}{i\gamma\cdot p' + m_e}\gamma_\lambda(1+\gamma_5)v(k_2) \times [\bar{u}(k_1)\gamma_\lambda(1+\gamma_5)u(p)]\cdot N\,, \qquad (14b)$$

$$T_{1c} = -\frac{G}{\sqrt{2}}N\frac{ie^2}{(2\pi)^4}\bar{u}(p')\int\frac{d^4Q}{Q^2}\gamma_\nu\frac{1}{i\gamma\cdot(p'-Q)+m_e}\gamma_\lambda \times (1+\gamma_5)v(k_2)\bar{u}(k_1)\gamma_\lambda(1+\gamma_5)\frac{1}{i\gamma\cdot(p-Q)+m_\mu}\gamma_\nu u(p) \qquad (14c)$$

where the quantity

$$\Sigma^{(2)}(p) = \frac{ie^2}{(2\pi)^4} \int \frac{d^4Q}{Q^2} \gamma_\nu \frac{1}{i\gamma \cdot (p-Q) + m} \gamma_\nu \tag{14d}$$

appearing in T_{1a} and T_{1b} is the familiar second-order self-energy operator encountered in quantum electrodynamics and we deal with it as in that theory [9]. Thus we write $\Sigma^{(2)}(p)$ as

$$\Sigma^{(2)}(p) = A + (i\gamma \cdot p + m)B + (i\gamma \cdot p + m)^2 \Sigma^p \tag{15a}$$

where A and B turn to be logarithmically divergent constants but Σ^p is finite. The constant A is removed by mass renormalization (through the introduction of a counter-term in the interaction Hamiltonian). The constant B, which, neglecting its finite part can be written as

$$B = \frac{ie^2}{(2\pi)^4} \int \frac{d^4Q}{(Q^2+m^2)^2} = -\frac{\alpha}{4\pi} \ln\left(\frac{\Lambda^2}{m^2}\right) \tag{15b}$$

is connected with the wave function renormalization [9]. We note that in both T_{1a} and T_{1b}, $\Sigma^{(2)}(p)$ is multiplied by the factor $(i\gamma p + m)^{-1}$ and the combination acts upon the free spinor $u(p)$ or $\bar{u}(p)$; this makes the contribution of the term in $\Sigma^{(2)}(p)$ involving B ambiguous (since it involves the quotient %), but a careful treatment of this question shows that the correct value is $(B/2)$. The contribution of the Σ^p term in $\Sigma^{(2)}(p)$ to T_{1a} and T_{1b} is zero. Thus we see that

$$(T_{1a} + T_{1b})^{\text{divergent}} = -\frac{G}{\sqrt{2}} \left[\frac{ie^2}{(2\pi)^4} \int \frac{d^4Q}{Q^4}\right] \times [\bar{u}(k_1)\gamma_\lambda(1+\gamma_5)u(p)][\bar{u}(p')\gamma_\lambda(1+\gamma_5)v(k_2)] \cdot N \tag{16a}$$

After some manipulation and using the identity

$$\gamma_\lambda \gamma_\varrho \gamma_\nu = \delta_{\lambda\varrho}\gamma_\nu - \delta_{\lambda\nu}\gamma_\varrho + \delta_{\varrho\nu}\gamma_\lambda + \varepsilon_{\lambda\varrho\nu\sigma}\gamma_5\gamma_\sigma$$

one can write the divergent part of T_{1c} as

$$T_{1c}^{\text{divergent}} = -\frac{G}{\sqrt{2}} \left[\frac{-ie^2}{(2\pi)^4} \int \frac{d^4Q}{Q^4}\right] \times [\bar{u}(k_1)\gamma_\lambda(1+\gamma_5)u(p)][\bar{u}(p')\gamma_\lambda(1+\gamma_5)v(k_2)] \cdot N . \tag{16b}$$

The importance of the foregoing discussion is, as we can see by inspection of Eqs. (16a, b), that in the sum $T_1 = T_{1a} + T_{1b} + T_{1c}$ the various divergences cancel. Hence the second-order radiative corrections to μ-decay are free of ultraviolet divergences.

A circumstance such as the present cancellation of divergences has a more direct explanation [10]. We recall that a similar cancellation occurs [9] in quantum electrodynamics where the divergences from the three Feynman diagrams shown in Fig. 2 cancel. The reason for the cancellation there is precisely the form of the interaction

$$\mathcal{L}_{\text{int}}^{\text{e.m.}} = -ej_\lambda^{\text{e.m.}}(x)a_\lambda(x) = -ie[\bar{l}(x)\ \gamma_\lambda l(x)]a_\lambda(x) \tag{17}$$

where $l=e^-$ or μ^- and $j_\lambda^{\text{e.m.}} = i\bar{l}\gamma_\lambda l$ is a divergenceless neutral vector current interacting with the electromagnetic potential a_λ (neutral vector field). With the interaction (17), the Ward identity ensures a cancellation of the vertex corrections (Fig. 2c) and the self-energy parts (Figs. 2a, b)

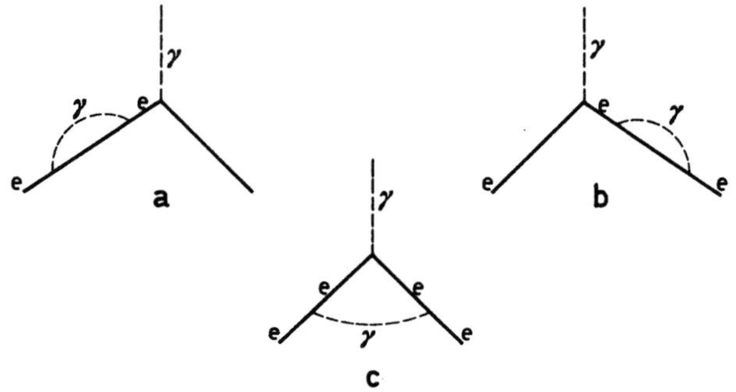

Fig. 2. Feynman diagrams for radiative corrections to electromagnetic interactions

after mass renormalization. Now the interaction (12) for μ-decay can be rewritten by *Fierz* re-ordering in the form

$$\mathcal{L}_{\text{W}}^{\mu} = \frac{G}{\sqrt{2}}[i\bar{e}(x)\gamma_\lambda(1+\gamma_5)\mu(x)][i\bar{\nu}_\mu(x)\gamma_\lambda(1+\gamma_5)\nu_e(x)] + \text{h.c.} \tag{18}$$

The interaction (18) also has the form of a neutral current $i\bar{e}\gamma_\lambda(1+\gamma_5)\mu$ made up of charged fields interacting with a neutral vector field, i.e. $i[\bar{\nu}_\mu(x)\gamma_\lambda(1+\gamma_5)\nu_e(x)]$ with the following differences, however: (i) the term $i\bar{e}\gamma_\lambda\mu$ differs from the case of electrodynamics only in that μ and e have different masses (i.e. the current $i\bar{e}\gamma_\lambda\mu$ is not divergence free), (ii) there is also an axial vector part $i\bar{e}\gamma_\lambda\gamma_5\mu$. With respect to (i), since the ultraviolet divergences are only logarithmic (and therefore do not really depend on mass since one can write $\ln(\Lambda/m') = \ln(\Lambda/m) +$ finite part $\ln m/m'$) both from the vertex part and self-energy part (after mass renormalization) they still cancel. What remains is a finite contribution depending upon the masses, which would also cancel, due to the Ward identity in the limit $m_\mu = m_e$. Concerning (ii) we note that the axial vector interaction becomes equivalent to the vector interaction through the replacement $\mu \to \gamma_5\mu$, $m_\mu \to -m_\mu$. Therefore, in this case also the ultraviolett divergences cancel, while there remains a finite part which cancels for $m_e = -m_\mu$. Thus the radiative corrections to μ-decay are free of ultraviolet divergences.

That the above cancellation of divergences occurs depends sensitively on the form of the interaction (12) or (18). For instance, for the case of neutron β-decay

$$n \to p + e^- + \bar{\nu}_e$$

where the effective phenomenological weak Lagrangian can be written as

$$-\frac{G}{\sqrt{2}} [\bar{p}\gamma_\lambda(1+\varrho\gamma_5)n][\bar{e}\gamma_\lambda(1+\gamma_5)\nu_e] + \text{h.c.} \tag{19}$$

where $\varrho = 1.2$ experimentally, the divergences do not cancel [10]. First, we can rewritte (19) as

$$-\frac{G}{\sqrt{2}} [\bar{n}^c\gamma_\lambda(1-\varrho\gamma_5)p^c][\bar{e}\gamma_\lambda(1+\gamma_5)\nu_e] + \text{h.c.} \tag{19a}$$

which, by Fierz re-arrangement can be put in the form

$$\begin{aligned} -\frac{G}{\sqrt{2}}\Bigg[&\left(\frac{1-\varrho}{2}\right)\{\bar{e}\gamma_\lambda(1+\gamma_5)p^c\}\{\bar{n}^c\gamma_\lambda(1+\gamma_5)\nu_e\} \\ &-\frac{1+\varrho}{2}\{\bar{e}(1-\gamma_5)p^c\}\{\bar{n}^c(1+\gamma_5)\nu_e\} + \text{h.c.}\Bigg]. \end{aligned} \tag{19b}$$

The argument based on electrodynamic analogy fails in the case of second term, this being of $S-P$ nature and the corrections to this term give rise to ultraviolet divergences. The radiative corrections to the neutron β-decay are thus in general divergent if we treat the nucleons as point-like particles and they converge only in the special case $\varrho = -1$, while experimentally $\varrho = 1.2$.

In order to see that the cancellation of divergences depends on the form of interaction in more general context, we compute in lowest-order perturbation theory the divergent parts of the radiative corrections to

$$x \to y + e^- + \bar{\nu}_e \tag{20}$$

where x and y stand for general spin 1/2 point-like particles having electric charges Q_x and $Q_y\,(=Q_x+1)$ and where the weak interaction is of $V-\varrho A$ form

$$-\frac{G}{\sqrt{2}} [\bar{y}\gamma_\lambda(1+\varrho\gamma_5)x][\bar{e}\gamma_\lambda(1+\gamma_5)\nu_e] + \text{h.c.} \tag{21}$$

In this situation, we have in addition to diagrams (a), (b), and (c) of Fig. 1 with μ and ν_μ replaced by x and y, the diagrams (d), (e), and (f)

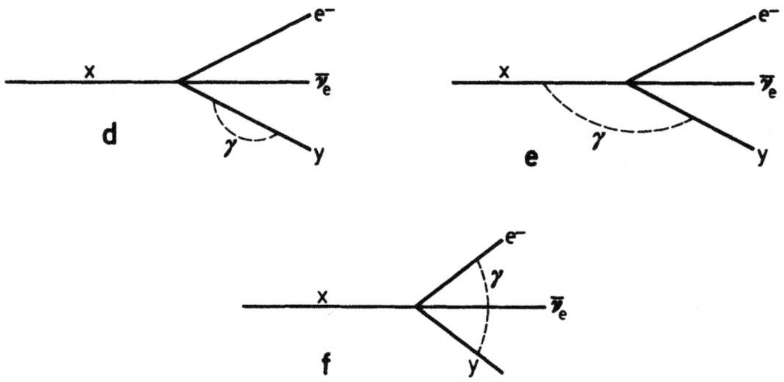

Fig. 3. Additional Feynman diagrams for radiative corrections to $x \to y + e^- + \bar{\nu}_e$

shown in Fig. 3. The divergent parts of diagrams (a), (b), and (d) are now given by

$$T_{1a,b,d}^{\text{divergent}} = -\frac{G}{\sqrt{2}} N \cdot \frac{1}{2}(Q_x^2, 1, Q_y^2) \times \frac{ie^2}{(2\pi)^4} \int \frac{d^4Q}{Q^4} \bar{u}_y \gamma_\lambda (1 + \varrho\gamma_5) u_x \bar{u}_e \gamma_\lambda (1 + \gamma_5) v_\nu \tag{22a}$$

while those of diagrams (c), (e) and (f) are, respectively,

$$T_{1c}^{\text{divergent}} = -\frac{G}{\sqrt{2}} N \cdot \frac{Q_x}{4} \frac{ie^2}{(2\pi)^4} \int \frac{d^4Q}{Q^4} \times \bar{u}_y \gamma_\lambda [(10 - 6\varrho) + (10\varrho - 6)\gamma_5] u_x \bar{u}_e \gamma_\lambda (1 + \gamma_5) v_\nu \,, \tag{22b}$$

$$T_{1e}^{\text{divergent}} = -\frac{G}{\sqrt{2}} N \cdot (-Q_x Q_y) \frac{ie^2}{(2\pi)^4} \int \frac{d^4Q}{Q^4} \times \bar{u}_y \gamma_\lambda (1 + \varrho\gamma_5) u_x \bar{u}_e \gamma_\lambda (1 + \gamma_5) v_\nu \,, \tag{22c}$$

$$T_{1f}^{\text{divergent}} = -\frac{G}{\sqrt{2}} N \cdot \left(-\frac{Q_y}{4}\right) \frac{ie^2}{(2\pi)^4} \int \frac{d^4Q}{Q^4} \times \bar{u}_y \gamma_\lambda [(10\varrho - 6) + (10 - 6\varrho)\gamma_5] u_x \bar{u}_e \gamma_\lambda (1 + \gamma_5) v_\nu \,. \tag{22d}$$

Upon adding up all these terms we obtain for the divergent part of the radiative corrections to the process (20) the expression

$$T_1^{\text{divergent}} = -\frac{G}{\sqrt{2}} N \cdot \left[\left(-\frac{3}{2} \frac{ie^2}{(2\pi)^4} \int \frac{d^4Q}{Q^4}\right] \times \{\bar{u}_y \gamma_\lambda [Q_y(1 + \varrho)(1 + \gamma_5) - Q_x(1 - \varrho)(1 - \gamma_5)] u_x\} \times \bar{u}_e \gamma_\lambda (1 + \gamma_5) v_\nu \,. \tag{23}$$

The divergence cancels if and only if

$$Q_y(1+\varrho)=0\,,\qquad Q_x(1-\varrho)=0\,. \tag{24}$$

These equations have two and only two solutions

$$Q_y=0,\quad Q_x=-1 \quad \text{and} \quad \varrho=1 \quad (V-A)\,,\qquad Q_x=0,\quad Q_y=1 \quad \text{and} \quad \varrho=-1(V+A)\,. \tag{25}$$

The first corresponds to the physical case of μ^--decay while the second corresponds to $n \to p+e^-+\bar{\nu}_e$ (n, p being point-like) with $V+A$ interaction. We also see that the radiative corrections are divergent in a bare $(V-\varrho A)$ quark model independent of ϱ since, in this case, $Q_x=-1/3$ and $Q_y=2/3$. The conditions (25) for the cancellation of divergences for point-like case will be appealed to in our later discussion.

Besides the ultraviolet divergences, the integrals in Eqs. (14) contain infrared divergences as well. These divergences are dealt with in the usual way [9], i.e. by adding to the probability for ordinary muon decay the probability for radiative decay involving a photon of energy below the detection efficiency of the experiment, say $(k_0)_{\min}$. Then to order α, the transition probability is given by

$$|T_0|^2+(T_0T_1^*+T_0^*T_1)+\int_0^{(k_0)_{\min}}|T_y|^2\,\mathrm{d}k_0$$

and in this sum the infrared divergences present in the second and third terms cancel one another. Hence the complete radiative corrections to muon decay are finite.

As regards the life-time of μ-decay, it has been shown that the corrected value $\tau^{(\mu)}$ after radiative corrections is given by

$$\tau^{(\mu)}=\tau_0^{(\mu)}\left[1+\frac{\alpha}{2\pi}\left(\pi^2-\frac{25}{4}\right)\right] \tag{26}$$

where τ_0^{μ} is the life-time in the absence of electromagnetic corrections. The radiative corrections increase the life-time by about 0.42%. If we define the observed μ-decay coupling G^{μ} by the relation

$$\Gamma^{(\mu)}=\frac{1}{\tau^{(\mu)}}=\frac{(G^{\mu})^2m_\mu^5}{192\pi^3} \tag{27}$$

and if G is the bare μ-decay constant so that

$$\Gamma^{(\mu)}=\frac{G^2m_\mu^5}{192\pi^3}\left[1-\frac{\alpha}{2\pi}\left(\pi^2-\frac{25}{4}\right)\right]. \tag{28}$$

We deduce that

$$G^{\mu} = G\left[1 - \frac{\alpha}{4\pi}\left(\pi^2 - \frac{25}{4}\right)\right] \tag{29a}$$

or

$$G = G^{\mu}\left[1 + \frac{\alpha}{4\pi}\left(\pi^2 - \frac{25}{4}\right)\right]. \tag{29b}$$

Thus the "bare" μ-decay coupling constant is larger than the observed μ-decay coupling constant by about 0.21%. Now

$$G^{\mu} = (1.4320 \pm 0.0011) \times 10^{-49}\ \text{erg cm}^3$$

as calculated from (27). Then (29b) gives

$$G = (1.4350 \pm 0.0011) \times 10^{-49}\ \text{erg cm}^3. \tag{30}$$

3. Radiative Corrections to $\Delta Y = 0$ Semi-leptonic Processes

a) Divergence Problems

The radiative corrections of order α to semi-leptonic weak processes of the type $A \to B + l + \nu_l$, which has the uncorrected matrix element

$$\begin{aligned} T_0 = i\frac{G^0}{\sqrt{2}} &\langle B(p')|J^0_\lambda(0)|A(p)\rangle \\ &\times \frac{1}{(2\pi)^3}\sqrt{\frac{m_\nu m_l}{k_0 k_0'}}\,\bar{u}(k')\gamma_\lambda(1+\gamma_5)v(k) \end{aligned} \tag{31}$$

arise in lowest order from the three Feynman diagrams in Fig. 4. We denote the reduced T-matrix elements corresponding to the Feynman diagrams 4(a) (hadronic), 4(b) (hadronic-leptonic) and 4(c) (leptonic),

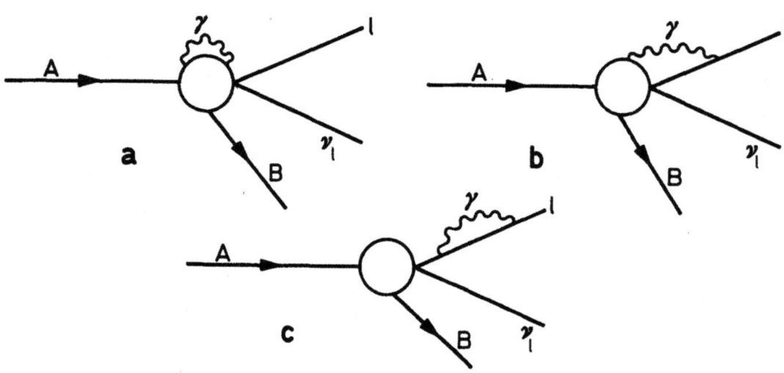

Fig. 4. Feynman diagrams for the radiative corrections to $A \to B + l + \nu_l$

respectively, by T_H, T_{HL} and T_L. We can write

$$T_H=\frac{iG^0}{\sqrt{2}}\int d^4Q\,\frac{-i}{(2\pi)^4}\,\frac{\delta_{\mu\nu}}{Q^2}\,\frac{1}{(2\pi)^3}\sqrt{\frac{m_\nu m_l}{k_0 k_0'}}\;u(k')\,\gamma_\lambda(1+\gamma_5)\,v(k) \qquad (32\text{a})$$
$$\times\langle B(p')\gamma(\mu,Q)\gamma(\nu,-Q)_{\text{out}}|J^0_\lambda(0)|A(p)\rangle$$

where, using the LSZ formalism,

$$\langle B(p')\gamma(\mu,Q)\gamma(\nu,-Q)_{\text{out}}|J^0_\lambda(0)|A(p)\rangle=\frac{i^2e^2}{2}\int d^4x\,d^4y\,e^{-iQ\cdot x}e^{iQ\cdot y}$$
$$\times\langle B(p')|T^*(j^{\text{e.m.}}_\mu(x)j^{\text{e.m.}}_\nu(y)J^0_\lambda(0))|A(p)\rangle=\frac{e^2}{2}\,\tilde{T}_{\mu\nu\lambda}(p,p',Q) \qquad (32\text{b})$$

with the factor 1/2 appearing to avoid counting the photon twice. Using translational invariance, we can also write $\tilde{T}_{\mu\nu\lambda}$ as [11]

$$\tilde{T}_{\mu\nu\lambda}(p,p',Q,q)=i^2\int d^4x\,d^4y\,e^{-iQ\cdot x}e^{iq\cdot y}$$
$$\times\langle B(p')|T^*(j^{\text{e.m.}}_\mu(x)j^{\text{e.m.}}_\nu(0)J^0_\lambda(y))|A(p)\rangle \qquad (32\text{c})$$

where $q=p'-p=-(k'+k)$. Hence

$$T_H=\frac{iG^0e^2}{2\sqrt{2}}\,\frac{-i}{(2\pi)^4}\int d^4Q\,\frac{\delta_{\mu\nu}}{Q^2}\,\frac{1}{(2\pi)^3}\sqrt{\frac{m_\nu m_l}{k_0k_0'}}$$
$$\times\bar{u}(k')\gamma_\lambda(1+\gamma_5)v(k)\,\tilde{T}_{\mu\nu\lambda}(p,p',Q,q)\,. \qquad (32\text{d})$$

Similarly, we obtain

$$T_{HL}=\frac{iG^0}{\sqrt{2}}\int d^4Q\,\frac{-i}{(2\pi)^4}\,\delta_{\mu\nu}\,\frac{1}{Q^2}\,\frac{ie}{(2\pi)^3}\sqrt{\frac{m_\nu m_l}{k_0k_0'}}$$
$$\times\bar{u}(k')\gamma_\nu\,\frac{1}{i\gamma\cdot(k'-Q)+m_l}\,\gamma_\lambda(1+\gamma_5)v(k) \qquad (33\text{a})$$
$$\times\langle B(p')\gamma(\mu,Q)_{\text{out}}|J^0_\lambda(0)|A(p)\rangle$$

where

$$\langle B(p')\gamma(\mu,Q)_{\text{out}}|J^0_\lambda(0)|A(p)\rangle$$
$$=ie\int d^4x\,e^{-iQ\cdot x}\langle B(p')|T^*(j^{\text{e.m.}}_\mu(x)J^0_\lambda(0))|A(p)\rangle \qquad (33\text{b})$$
$$=e\,T_{\mu\lambda}(p,p',Q)\,.$$

Thus

$$T_{HL}=\frac{iG^0e^2}{\sqrt{2}(2\pi)^4}\int\frac{d^4Q}{Q^2}\,T_{\mu\lambda}(p,p',Q)\,l_{\mu\lambda}\,, \qquad (33\text{c})$$

where

$$l_{\mu\lambda} = \frac{1}{(2\pi)^3}\sqrt{\frac{m_\nu m_l}{k_0 k_0'}}\,\bar{u}(k')\gamma_\mu \frac{1}{i\gamma\cdot(k'-Q)+m_l}\gamma_\lambda(1+\gamma_5)v(k)$$
$$= \frac{1}{(2\pi)^3}\sqrt{\frac{m_\nu m_l}{k_0 k_0'}}\frac{i}{Q^2-2k'\cdot Q}[-(Q+2k')_\mu\delta_{\lambda\lambda'} \tag{33d}$$
$$+Q_{\lambda'}\delta_{\lambda\mu}-Q_\lambda\delta_{\lambda'\mu}+\varepsilon_{\mu\varrho\lambda\lambda'}Q_\varrho]\bar{u}(k')\gamma_{\lambda'}(1+\gamma_5)v(k)\,.$$

The last result follows by using the properties of γ-matrices. Finally, by using standard perturbation theory for the leptonic part, we have for the Feynman diagram 4(c)

$$T_L = \frac{iG^0}{\sqrt{2}}\left[\frac{1}{(2\pi)^3}\sqrt{\frac{m_\nu m_l}{k_0 k_0'}}\,\bar{u}(k')\Sigma^{(2)}(k')\right.$$
$$\left.\times\frac{1}{i\gamma\cdot k'+m_l}\gamma_\lambda(1+\gamma_5)v(k)\right]\langle B(p')|J_\lambda^0(0)|A(p)\rangle \tag{34}$$

where $\Sigma^{(2)}(k')$ is given in (14d) and can be treated as before. After performing the mass renormalization, $\Sigma^{(2)}(k')$ has the effect of modifying the coupling constant G^0 in the uncorrected matrix element T_0 by the amount $(\delta G^0)_L = G^0(1/2B)$. This fact is denoted by

$$(\delta T^0)_L^{\text{div}} = T_L^{\text{div}} = \frac{iG^0}{\sqrt{2}}\left[\frac{1}{2}\frac{ie^2}{(2\pi)^4}\int\frac{d^4Q}{Q^4}\right]$$
$$\times\langle B(p')|J_\lambda^0(0)|A(p)\rangle\frac{1}{(2\pi)^3}\sqrt{\frac{m_\nu m_l}{k_0 k_0'}}\,\bar{u}(k')\gamma_\lambda(1+\gamma_5)v(k) \tag{35}$$

where the quantity within the square brackets corresponds to $(1/2B)$. Introducing the definitions

$$l_\lambda = \frac{1}{(2\pi)^3}\sqrt{\frac{m_\nu m_l}{k_0 k_0'}}\,\bar{u}(k')\gamma_\lambda(1+\gamma_5)v(k)\,,$$
$$\alpha = \frac{e^2}{4\pi}\,, \tag{36}$$

and

$$D = \frac{1}{i\pi^2}\int\frac{d^4Q}{Q^4} \tag{37}$$

for the logarithmically divergent integral, we can write (35) as

$$T_L^{\text{div}} = \frac{iG^0}{\sqrt{2}}\left(-\frac{\alpha}{8\pi}D\right)\langle B(p')|J_\lambda^0(0)|A(p)\rangle l_\lambda\,. \tag{38}$$

We would now like to inquire whether the divergent radiative correction contributed by T_L^{div} is cancelled by similar divergent contributions associated with T_H and T_{HL}, which is precisely what happened in the case of μ-decay.

Before proceeding to this question, we remark that in addition to the above ultraviolet divergences, the matrix elements T_H, T_{HL} and T_L also contain infrared divergences. The procedure for dealing with them has already been described.

We now discuss T_{HL} and T_H in turn [12], [13]. Using the expression (33b) for $T_{\mu\lambda}$, we note that

$$\begin{aligned} Q_\mu T_{\mu\lambda} &= \int \mathrm{d}^4x\, \mathrm{e}^{-iQ\cdot x} \langle B|\delta(x_0)[j_0^{\text{e.m.}}(x), J_\lambda^0(0)]|A\rangle \\ &= \int \mathrm{d}^4x\, \mathrm{e}^{-iQ\cdot x} \langle B|\delta(x_0)[V_{30}(x), J_{1+i2,\lambda}(0)]|A\rangle \\ &= \langle B|J_\lambda^0(0)|A\rangle \end{aligned} \tag{39}$$

where we have used $\partial_\mu j_\mu^{\text{e.m.}} = 0$ and the current commutation relations (2). Thus, using the expression (33d) for $l_{\mu\lambda}$, we can write T_{HL}

$$\begin{aligned} T_{HL} = \frac{G^0}{\sqrt{2}} \frac{e^2}{(2\pi)^4} \int \frac{\mathrm{d}^4Q}{Q^2(Q^2 - 2k'\cdot Q)} & \\ \times [\langle B|J_\lambda^0(0)|A\rangle + Q_\mu T_{\lambda\mu} - Q_\lambda T_{\mu\mu} & \\ + 2k'_\mu T_{\mu\lambda} - \varepsilon_{\mu\varrho\nu\lambda} Q_\varrho T_{\mu\nu}] l_\lambda \,. & \end{aligned} \tag{40}$$

We next show that $\tilde{T}_{\mu\nu\lambda}$, which enters T_H, can be related to $T_{\mu\lambda}$ by using the algebra of currents. For this purpose we treat q in the expression (32c) for $\tilde{T}_{\mu\nu\lambda}(p, p', Q, q)$ as an independent variable which we put equal to $p' - p$ at the end. Then, using the identity

$$\begin{aligned} -iq_\lambda \tilde{T}_{\mu\nu\lambda}(p, p', Q, q) &= \tilde{M}_{\mu\lambda}(p, p', Q, q) + i^2 \int \mathrm{d}^4x\, \mathrm{d}^4y\, \mathrm{e}^{-iQ\cdot x} \mathrm{e}^{iq\cdot y} \\ &\times [\delta(y_0)\langle B|T\{j_\mu^{\text{e.m.}}(x)[J_0^0(y), j_\nu^{\text{e.m.}}(0)]\}|A\rangle \\ &+ \delta(x_0 - y_0)\langle B|T\{j_\nu^{\text{e.m.}}(0)\,[J_0^0(y), j_\mu^{\text{e.m.}}(x)]\}|A\rangle] \,, \end{aligned} \tag{41}$$

where

$$\begin{aligned} \tilde{M}_{\mu\nu}(p, p', Q, q) = i^2 \int \mathrm{d}^4x\, \mathrm{d}^4y\, \mathrm{e}^{-iQ\cdot x} \mathrm{e}^{iq\cdot y} & \\ \times \langle B(p')|\,\tilde{T}(j_\mu^{\text{e.m.}}(x) j_\nu^{\text{e.m.}}(0)\partial_\lambda J_\lambda^0(y))|A(p)\rangle \,, & \end{aligned} \tag{42}$$

and the current commutation relations

$$\begin{aligned} \delta(x_0 - y_0)[J_0^0(y), j_\mu^{\text{e.m.}}(x)] & \\ = \delta(x_0 - y_0)[J_{1+i2,0}(y), V_{3\mu}(x)] = -J_\mu^0(x)\delta^4(x - y) \,, & \end{aligned} \tag{43}$$

which follow from (2), we obtain on using the translational invariance [14]

$$-iq_\varrho \tilde{T}_{\mu\nu\varrho}(p,p',Q,q) \tag{44}$$
$$= -iT_{\mu\nu}(p,p',Q) - iT_{\nu\mu}(p,p',-Q+q-p'+p) + \tilde{M}_{\mu\nu}(p,p',Q,q).$$

It is easy to see [15] that $\tilde{T}_{\mu\nu\lambda}(p,p',Q,q)$ has poles at $q=p'-p$. These poles arise from one-particle intermediate states and correspond to the self-energy insertions in the external lines A and B as shown in

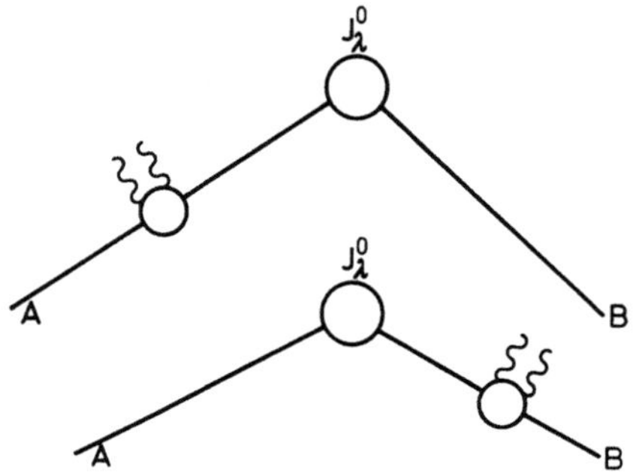

Fig. 5. Feynman diagrams for self-energy insertions in the external lines

Fig. 5. We remove these poles simply by subtracting them out, i.e. we replace $\tilde{T}_{\mu\nu\lambda}$ by

$$T_{\mu\nu\lambda} = \tilde{T}_{\mu\nu\lambda} - \delta M_{\mu\nu\lambda} \tag{45}$$

where $\delta M_{\mu\nu\lambda}$ is a tensor such that when inserted in Eq. (32d) it gives the electromagnetic contribution to the mass of the initial and final hadrons A and B; this corresponds to mass renormalization by adding appropriate counter-terms in $H_{\text{int}}^{\text{e.m.}}$ corresponding to the electromagnetic self-masses of the particles. These counter-terms cancel the above poles at $q=p'-p$, so that $T_{\mu\nu\lambda}(p,p',Q,q) = [\tilde{T}_{\mu\nu\lambda}(p,p',Q,q) - \delta M_{\mu\nu\lambda}]$ and

$$M_{\mu\nu}(p,p',Q,q) = [\tilde{M}_{\mu\nu}(p,p',Q,q) - (-iq_\lambda \delta M_{\mu\nu\lambda})] \tag{46}$$

no longer have poles. Thus we rewrite (32d) and (44) respectively as

$$T_H = \frac{iG^0 e^2}{2\sqrt{2}} \frac{-i}{(2\pi)^4} \int d^4Q \frac{\delta_{\mu\nu}}{Q^2} T_{\mu\nu\lambda}(p,p',Q,q) l_\lambda, \tag{47}$$

$$-iq_\varrho T_{\mu\nu\varrho}(p,p',Q,q) = -iT_{\mu\nu}(p,p',Q) - iT_{\nu\mu}(p,p',-Q+q-p'+p) + M_{\mu\nu}(p,p',Q,q). \tag{48}$$

We can now differentiate (48) with respect to q_λ, then put $q=p'-p=-(k'+k)$ so as to obtain

$$T_{\mu\nu\lambda}(p,p',Q,q=p'-p)=\frac{\partial}{\partial q_\lambda}T_{\nu\mu}(p,p',-Q+q-p'+p)|_{q=p'-p}$$
$$-(p'-p)_\varrho\frac{\partial}{\partial q_\lambda}T_{\mu\nu\varrho}(p,p',Q,q)|_{q=p'-p}+i\frac{\partial}{\partial q_\lambda}M_{\mu\nu}(p,p',Q,q)|_{q=p'-p}\,. \tag{49}$$

From the expression (33) for $T_{\nu\mu}$, it follows that the first term on the r.h.s. of (49) can be written as

$$-\frac{\partial}{\partial Q_\lambda}T_{\nu\mu}(p,p',-Q)\,. \tag{50}$$

Hence, substituting (49) in (46) and performing a partial integration on the term (50) we obtain

$$T_H=\frac{G^0}{\sqrt{2}}\frac{e^2}{(2\pi)^4}\int\frac{d^4Q}{Q^2}\left[\frac{1}{Q^2}T_{\mu\mu}(p,p',Q)Q_\lambda\right.$$
$$+\frac{i}{2}\frac{\partial}{\partial q_\lambda}M_{\mu\mu}(p,p',Q,q)|_{q=p'-p} \tag{51}$$
$$\left.-\frac{1}{2}(p'-p)_\varrho\frac{\partial}{\partial q_\lambda}T_{\mu\mu\varrho}(p,p',Q,q)|_{q=p'-p}\right]l_\lambda\,.$$

Two remarks are in order: (i) in view of our discussion following Eq. (44) it is seen that $T_{\mu\nu\lambda}$ has no singularity at $q=0$ so that the last term on the r.h.s. of (51) is of order $(p'-p)$. This means that if A and B are members of an isospin multiplet, the last term would be of order α^2 since then $(p'-p)\sim\alpha$; (ii) if the current J^0_λ is conserved, the second term on the r.h.s. of (51) is zero (see Eq. (42)). Thus, for the case of a conserved current and A and B belonging to an isospin multiplet, only the first term on the r.h.s. of (51) contributes to order α. We shall make use of this result shortly.

Adding the hadronic-leptonic contribution (40) to (51), we obtain

$$(\delta T_0)_{H+HL}=T_H+T_{HL}=\frac{G^0}{\sqrt{2}}\frac{e^2}{(2\pi)^4}\int\frac{d^4Q}{Q^2(Q^2-2k'\cdot Q)}$$
$$\times\left[\langle B|J^0_\lambda(0)|A\rangle+Q_\mu T_{\lambda\mu}(p,p',Q)\right. \tag{52}$$
$$-\frac{2k'\cdot Q}{Q^2}Q_\lambda T_{\mu\mu}(p,p',Q)+2k'_\mu T_{\mu\lambda}(p,p',Q)$$
$$\left.+\varepsilon_{\varrho\mu\nu\lambda}Q_\varrho T_{\mu\nu}\right]l_\lambda+\frac{G^0}{\sqrt{2}}(I^M_\lambda+I^T_\lambda)l_\lambda$$

where $(I_\lambda^M + I_\lambda^T)$ denotes the contribution of the last two terms on the r.h.s. of (51).

We see that the first term within the square brackets of Eq. (52) gives a logarithmically divergent contribution. In order to discuss the convergence or divergence of the other terms within the square brackets in (52), we need to know the asymptotic behaviour of $T_{\mu\lambda}$; at this stage, we invoke the Bjorken technique (cf. Eq. (9)) which permits us to write

$$\underset{\boldsymbol{Q}=0}{T_{\mu\lambda}} \xrightarrow[Q_0\to\infty]{} \frac{-1}{Q_0} \int d^3x \langle B|[j_\mu^{\text{e.m.}}(\boldsymbol{x},0),\, J_\lambda^0(0)]|A\rangle \tag{53}$$

where the frame $\boldsymbol{Q}=0$ is chosen in order to avoid Schwinger terms which might appear in the equal-time commutator in (53). We thus see that the third and fourth terms on the r.h.s. of (52) give convergent contributions. It follows from (53) that the second and fifth terms, which give divergent contributions, have the asymptotic behaviour

$$\begin{aligned}\lim_{\substack{Q_0\to\infty\\ \boldsymbol{Q}=0}} &[Q_\mu T_{\lambda\mu} + \varepsilon_{\varrho\mu\nu\lambda} Q_\varrho T_{\mu\nu}]\\ &= -\frac{1}{Q_0}[Q_4 \langle B|[\textstyle\int j_\lambda^{\text{e.m.}}(\boldsymbol{x},0)d^3x,\, J_4^0(0)]|A\rangle \\ &\quad + Q_4 \varepsilon_{4lm\lambda} \langle B|[\textstyle\int j_l^{\text{e.m.}}(\boldsymbol{x},0)d^3x,\, J_m^0(0)]|A\rangle]\,.\end{aligned} \tag{54}$$

The first term on the right-hand side of (54) is determined by the time-space commutation relation (43) and is given by

$$-\frac{1}{Q_0}[iQ_4 \langle B|J_\lambda^0(0)|A\rangle]\,. \tag{55}$$

The second term on the r.h.s. of (54) is determined by the space-space commutation relations

$$[\textstyle\int j_l^{\text{e.m.}}(\boldsymbol{x},0)d^3x,\, J_m^0(0)] \tag{56}$$

which are model dependent. In the field algebra [16] this commutator is zero. In triplet models, where vector and axial vector currents are given by

$$V_{i\lambda} = i\bar{\psi}(x)\gamma_\lambda \frac{\lambda_i}{2}\psi(x)\,,$$

$$A_{i\lambda} = i\bar{\psi}(x)\gamma_\lambda\gamma_5 \frac{\lambda_i}{2}\psi(x)$$

where $\psi(x)$ is a unitary triplet, the space-space commutation relations are given by

$$[V_{il}(x), V_{jm}(y)]_{x_0=y_0} = [i\varepsilon_{lmn}d_{ijk}A_{kn}(x) + \delta_{lm}f_{ijk}V_{k4}(x)]\delta^3(\boldsymbol{x}-\boldsymbol{y}), \quad (57\text{a})$$

$$[V_{il}(x), A_{jm}(y)]_{x_0=y_0} = [i\varepsilon_{lmn}d_{ijk}V_{kn}(x) + \delta_{lm}f_{ijk}A_{k4}(x)]\delta^3(\boldsymbol{x}-\boldsymbol{y}), \quad (57\text{b})$$

$$[A_{il}(x), A_{jm}(y)]_{x_0=y_0} = [i\varepsilon_{lmn}d_{ijk}A_{kn}(x) + \delta_{lm}f_{ijk}V_{kn}(x)]\delta^3(\boldsymbol{x}-\boldsymbol{y}) \quad (57\text{c})$$

and the electromagnetic current is model dependent and can be written in general as

$$j_\lambda^{\text{e.m.}} = a_1 V_{3\lambda} + a_2 \frac{1}{\sqrt{3}} V_{8\lambda} + a_3 \sqrt{\frac{2}{3}}\, V_{0\lambda} \quad (58)$$

where a_1, a_2, a_3 are constants which depend on the model. Then from (57) and (58) it follows that [17]

$$\begin{aligned}[j_l^{\text{e.m.}}(\boldsymbol{x},0), J_m^0(0)] &= [j_l^{\text{e.m.}}(\boldsymbol{x},0), V_{1+i2m}(0) + A_{1+i2m}(0)] \\ &= -i\delta_{lm}(V_{1+i24}(0) + A_{1+i24}(0))\delta^3(\boldsymbol{x}) \\ &\quad + i\varepsilon_{lmn}\left(\frac{a_2}{3} + \frac{2a_3}{3}\right)(A_{1+i2n}(0) + V_{1+i2n}(0))\delta^3(\boldsymbol{x}).\end{aligned} \quad (59)$$

The situation regarding the constants a_1, a_2, and a_3 for the various triplet models is summarized in Table 1.

Thus from (55) and (59) we can write (54) as

$$\begin{aligned}\lim_{\substack{Q_0\to\infty \\ \boldsymbol{Q}=0}} (Q_\mu T_{\lambda\mu} + \varepsilon_{\varrho\mu\nu\lambda} Q_\varrho T_{\mu\nu}] &= -\frac{1}{Q_0}\Big\{iQ_4\langle B|J_\lambda^0(0)|A\rangle + Q_4\varepsilon_{4lm\lambda}\frac{i}{3}(a_2+2a_3)\varepsilon_{lmn4} \\ &\quad \times [\langle B|A_n^0(0)|A\rangle + \langle B|V_n^0(0)|A\rangle]\Big\}, \quad \text{for triplet models} \\ &= -\frac{1}{Q_0}[iQ_4\langle B|J_\lambda^0|A\rangle], \quad \text{for field algebra.}\end{aligned}$$

(60a), (60b)

The covariant transcription of this limit is straightforward:

$$\begin{aligned}[Q_\mu T_{\lambda\mu} + \varepsilon_{\varrho\mu\nu\lambda}Q_\varrho T_{\mu\nu}] &\to \langle B|J_\lambda^0(0)|A\rangle - \frac{1}{3}(a_2+2a_3) \\ &\quad \times \varepsilon_{\alpha\mu\nu\lambda}\varepsilon_{\mu\nu\varrho\beta}\frac{Q_\alpha Q_\beta}{Q^2}[\langle B|A_\varrho^0(0)|A\rangle + \langle B|V_\varrho^0(0)|A\rangle], \text{ for triplet models} \quad (61\text{a}) \\ &\to \langle B|J_\lambda^0(0)|A\rangle, \text{ for field algebra.} \quad (61\text{b})\end{aligned}$$

Table 1. *Electromagnetic structure constants in various triplet models*

Model	Triplet	$j_\lambda^{\text{e.m.}}$	a_1	a_2	a_3
Quark [18]	$I=\frac{1}{2}$ 0 $Q=\frac{2}{3},\ -\frac{1}{3}\ -\frac{1}{3}$	$i\left[\frac{2}{3}\bar{\psi}_1\gamma_\lambda\psi_1-\frac{1}{3}\bar{\psi}_2\gamma_\lambda\psi_2-\frac{1}{3}\bar{\psi}_3\gamma_\lambda\psi_3\right]$ $=\frac{i}{2}\bar{\psi}\gamma_\lambda\left[\lambda_3+\frac{1}{\sqrt{3}}\lambda_8\right]\psi$	1	1	0
Sakata [19]	$Q=1,\quad 0\quad 0$	$i\bar{\psi}_1\gamma_\lambda\psi_1=\frac{i}{2}\bar{\psi}\gamma_\lambda\left[\lambda_3+\frac{1}{\sqrt{3}}\lambda_8+\sqrt{\frac{2}{3}}\lambda_0\right]\psi$	1	1	1
Maki-Hara [20] "Ξ^0" "Ξ^-" "Λ^0"	$Q=0,\quad -1\quad 0$	$-i\bar{\psi}_2\gamma_\lambda\psi_2=-\frac{i}{2}\bar{\psi}\gamma_\lambda\times\left[-\lambda_3+\frac{1}{\sqrt{3}}\lambda_8+\sqrt{\frac{2}{3}}\lambda_0\right]\psi$	1	−1	−1
"Ξ^0" "Ξ^-" "Ω^-"	$Q=0,\quad -1\quad -1$	$-i[\bar{\psi}_2\gamma_\lambda\psi_2+\bar{\psi}_3\gamma_\lambda\psi_3]$ $=\frac{i}{2}\bar{\psi}\gamma_\lambda\left[\lambda_3+\frac{1}{\sqrt{3}}\lambda_8-2\sqrt{\frac{2}{3}}\lambda_0\right]\psi$	1	1	−2

Thus, from (35), (52) and (61) and noting that

$$\int\frac{d^4Q}{Q^4}\varepsilon_{\alpha\mu\nu\lambda}\varepsilon_{\mu\nu\varrho\beta}\frac{Q_\alpha Q_\beta}{Q^2}=\frac{1}{4}\varepsilon_{\alpha\mu\nu\lambda}\varepsilon_{\mu\nu\varrho\alpha}\int\frac{d^4Q}{Q^4}$$
$$=-\frac{3}{2}\delta_{\lambda\varrho}\int d^4Q/Q^4$$

it follows that the divergent correction to T_0 is

$$(\delta T_0)_{\text{divergent}}=(T_L+T_H+T_{HL})_{\text{divergent}}=\frac{G^0}{\sqrt{2}}\left(\frac{e^2}{(2\pi)^4}\int\frac{d^4Q}{Q^4}\right)$$
$$\times\left\{\frac{3}{2}\langle B|J_\lambda^0|A\rangle-\frac{3}{2}c[\langle B|A_\lambda^0|A\rangle+\langle B|V_\lambda^0|A\rangle]\right\}l_\lambda \tag{62}$$
$$+\frac{G^0}{\sqrt{2}}(I_\lambda^M+I_\lambda^T)_{\text{divergent}}l_\lambda$$

where

$$c=-(1/3)(a_2+2a_3)\quad\text{for the triplet models}$$
$$=0\quad\text{for the field algebra.} \tag{62a}$$

In general we cannot say anything about I_λ^M and I_λ^T (and in particular for $(I_\lambda^M + I_\lambda^T)_{\text{div}}$) since they depend on the details of the strong interactions. This fact makes the application of the above discussion useful only in those cases when: (i) J_λ^0 in T_H (and therefore in I_λ^M) is conserved so that I_λ^M drops out, (ii) A and B belong to the same isospin multiplet, since then I_λ^T is of order α^2 and can therefore also be dropped. Thus we confine outselves to the radiative corrections to the matrix elements of the vector current V_λ^0 between members of an isospin multiplet. Important examples of this kind are $\pi^- \to \pi^0 + e^- + \bar{\nu}_e, O^{14} \to N^{14} + e^+ + \nu_e$, where only the vector current appears in T_0, the uncorrected matrix element. For cases like $n \to p + e + \tilde{\nu}_e$ and $\Sigma^- \to \Sigma^0 + e^- + \tilde{\nu}_e$, our remarks apply to the vector "charge" form factor (i.e. the coefficient of γ_μ). In all such cases, only those parts of T_L and T_H corresponding to the vector term in J_λ^0 contribute, while in $\tilde{T}_{LH}$, there are contributions from both the vector and axial vector parts of J_λ^0. Thus (52) and (62), respectively, become

$$(\delta T_0^V)_{H+HL} = \frac{G^0}{\sqrt{2}} \frac{e^2}{(2\pi)^4} \int \frac{d^4Q}{Q^2(Q^2 - 2k' \cdot Q)}$$
$$\times \left[\langle B | V_\lambda^0 | A \rangle + Q_\mu T_{\lambda\mu}^V(p, p', Q) \right. \tag{63}$$
$$\left. - \frac{2k' \cdot Q}{Q^2} Q_\lambda T_{\mu\mu}^V(p, p', Q) + 2k'_\mu T_{\mu\lambda}^V(p, p', Q) + \varepsilon_{\varrho\mu\nu\lambda} Q_\varrho T_{\mu\nu}^A \right] l_\lambda$$

$$(\delta T_0^V)_{\text{div}} = \frac{G^0}{\sqrt{2}} \left(\frac{e^2}{(2\pi)^4} \int \frac{d^4Q}{Q^4} \right)$$
$$\times \left[\frac{3}{2} \langle B | V_\lambda^0 | A \rangle - \frac{3}{2} c \langle B | V_\lambda^0 | A \rangle \right] l_\lambda \tag{64}$$
$$= \left[\frac{-ie^2}{(2\pi)^4} \int \frac{d^4Q}{Q^4} \right] \frac{3}{2} (1 - c) T_0^V.$$

Or, in other words, the total divergent correction to the vector coupling constant G^0 becomes

$$(\delta G^0)_{\text{div}} = \frac{3\alpha}{8\pi} D(1 - c) G^0 . \tag{65}$$

b) Elimination of Divergences in Certain Models

Eq. (64) is our chief result, i.e. the electromagnetic radiative correction to the β-decay vector matrix element is, in general, logarithmically divergent independent of the details of strong interactions. The divergence is, of course, model-dependent because of the presence of the model-dependent constant c. This model-dependent contribution arises, as we saw, from the axial vector contribution to T_{HL} where one has to use the space-space commutation relations between $j_\lambda^{\text{e.m.}}$ and J_μ^0. On the other hand, the divergent contribution of T_H is model-independent because it involves the vector $\Delta Y=0$ hadron current and only the time-time and time-space commutation relations were employed. We see that the divergence appearing in (64) disappears if $c=1$. This condition cannot be achieved in field algebra which requires $c=0$. In certain triplet models mentioned in Table 1, however, it is possible to secure [13] $c=1$.

Let us now inquire what the condition $c=-1/3\,(a_2+2a_3)=1$ for the cancellation of divergences implies for the possible underlying triplet models. For such models, $j_\lambda^{\text{e.m.}}$ and J_λ^0 have, respectively, the forms

$$j_\lambda^{\text{e.m.}}=\frac{i}{2}\,\bar{\psi}\gamma_\lambda\left[a_1\lambda_3+\frac{1}{\sqrt{3}}\,a_2\lambda_8+\sqrt{\frac{2}{3}}\,a_3\lambda_0\right]\psi\,,$$

$$J_\lambda^0=i\bar{\psi}\,\frac{\lambda_1+i\lambda_2}{2}\,\gamma_\lambda(1+\gamma_5)\psi\,.$$

First we see from the commutation relation

$$[Q,J_\lambda^0]=J_\lambda^0,$$

where $Q=-i\int j_4^{\text{e.m.}}(\boldsymbol{x},0)\mathrm{d}^3x$ is the electric charge, that a_1 must be equal to 1. If we now also use the GNN relation

$$Q=I_3+Y/2\,, \tag{66a}$$

we see that the hypercharges of the triplet are given by

$$\begin{aligned} Y&=a_2\frac{1}{\sqrt{3}}\,\lambda_8+a_3\sqrt{\frac{2}{3}}\,\lambda_0\\ &=\frac{1}{3}\begin{pmatrix} a_2+2a_3 & & \\ & a_2+2a_3 & \\ & & 2(a_3-a_2)\end{pmatrix}. \end{aligned}\tag{66b}$$

Thus the constant $c=-Y_D$, where Y_D denotes the hypercharge of the isospin doublet belonging to the triplet ψ_i. Since I_3 for the doublet is $\pm 1/2$, we have

$$c=-Y_D=-2\bar{Q}$$

where $\bar{Q}$ is the average charge of the fundamental doublet (with $(V-A)$ structure in its β-decay) which underlies the current algebra.

Hence, we can write (65) as

$$(\delta G^0)_{\rm div} = G^0 \frac{3\alpha}{8\pi}(1+2\bar{Q})D \tag{67}$$

and the condition for the cancellation of divergence becomes

$$Y_D = -1 \quad \text{or} \quad \bar{Q} = -1/2\,. \tag{68}$$

From the GNN relation, this means that the charges of the isospin doublet ψ_1 and ψ_2 must be 0 and -1 respectively. In other words, in order that the divergences in (67) should cancel, the charge distribution for the isospin members of the fundamental triplet is that of the Ξ doublet Ξ^0 and Ξ^-. Purely from $\Delta Y = 0$ processes, one cannot say anything about the third member of the triplet. The possible triplets given in Table 1 which satisfy the above conditions are ("Ξ^0", "Ξ^-", "Λ^0") and ("Ξ^0", "Ξ^-", "Ω^-"). It is clear that the "fractional" quark model cannot give the desired cancellation of the divergences appearing in (68) and neither can the Sakata model. Note, however, that all of the foregoing discussion assumed that the current J_λ^0 is given by $(V_\lambda^0 + A_\lambda^0)$ with V_λ^0 and A_λ^0 obeying the equal-time commutation relations of Eqs. (2) and (57). If we were instead to assume that J_λ^0 is given by $(V_\lambda^0 - A_\lambda^0)$ with V_λ^0 and A_λ^0 obeying the same equal-time commutation relations, the only change in the above discussion would be the replacement of the factor $(1-c)$ appearing in $(\delta G^0)_{\rm div}$ (Eq. (67)) by the factor $(1+c)$ with c still given by (62a). In this second case, the condition for cancellation of divergences is $c=-1$ and this yields

$$Y_D = 1 \quad \text{or} \quad \bar{Q} = 1/2 \tag{69}$$

instead of (68). Eq. (69) implies that the charges of the isospinor members of the triplet are 1 and 0, as in the case of the nucleon doublet (p, n). In this case only the Sakata model among those listed in Table 1 gives finite radiative corrections. At the level of fundamental triplets, we may therefore summarize the position regarding the cancellation of divergences as follows. If the β-decay of the isodoublet members of the triplet is of the $(V-A)$ form (this corresponds to the current $J_\lambda^0 = V_\lambda^0 + A_\lambda^0$), then these members must have the charge distribution $(0, -1)$ (Ξ doublet) in order for the divergences in the radiative corrections to cancel. If, on the other hand, the β-decay of the isodoublet members of the triplet is of the $(V+A)$ form (this corresponds to the current $J_\lambda^0 = V_\lambda^0 - A_\lambda^0$) then, for the cancellation of the divergences it is necessary that the isodoublet possess the charges $(1, 0)$ (N doublet). It follows that the finiteness of the radiative corrections to the β-decay vector matrix element is strictly model-dependent. These conclusions are not surprising if one recalls the discussion of the radiative corrections to point-like particles. In the case of μ-decay we saw that for a muon current in the $(V-A)$ form, the radiative corrections were finite only if the charges of "μ" and "ν_μ" were -1 and $0((Q=-1)\,\mu^- \to (Q=0)\,\nu_\mu$ transition) respectively. On the other hand, for the $(V+A)$ form of the currents, the transition for which the radiative corrections are finite was found to be the $(Q=0)\to(Q=+1)$ transition. These correspond to ("Ξ^0", "Ξ^-") and ("p", "n") cases to which we have made reference.

c) Numerical Results

We now turn to a consideration of the finite parts of the radiative corrections. We shall first prove the following result [21]: if we examine the corrections to the vector matrix element arising from the electromagnetic corrections to the vector current only (i.e. put $A^0_\lambda = 0$ in T_{HL}) then we find that the finite contributions cancel out in the limit of zero four-momenta of the leptons. This can be seen by returning to the formula (63) where we put $T^A_{\mu\nu}=0$ and use the result

$$(Q+p'-p)_\mu T^V_{\lambda\mu}(p,p',Q) = \langle B|V^0_\lambda|A\rangle \tag{70}$$

which follows from the current commutation relations (43) and $\partial_\mu V^0_\mu = 0$. We thus obtain

$$\begin{aligned} T^V_H + T^V_{HL} = \frac{G^0}{\sqrt{2}}\frac{e^2}{(2\pi)^4}\int &\frac{d^4Q}{Q^2(Q^2-2k'\cdot Q)} \\ &\times\Big[2\langle B|V^0_\lambda|A\rangle + (k+k')_\mu T^V_{\lambda\mu} \\ &-\frac{2k'\cdot Q}{Q^2}Q_\lambda T^V_{\mu\mu} + 2k'_\mu T^V_{\mu\lambda}(p,p',Q)\Big]\, l_\lambda\,. \end{aligned} \tag{71 a}$$

By setting the lepton momenta equal to zero, we immediately get

$$\begin{aligned} (T^V_H + T^V_{HL})_{k'=0} &= \frac{G^0}{\sqrt{2}}\left(\frac{e^2}{(2\pi)^4}\int\frac{d^4Q}{Q^4}\right)2\langle B|V^0_\lambda|A\rangle l_\lambda \\ &= \frac{iG^0}{\sqrt{2}}\langle B|V^0_\lambda|A\rangle l_\lambda \frac{2\alpha}{4\pi}D\,. \end{aligned} \tag{71 b}$$

Finally, adding the correction to G^0 from T_L, one obtains for the total correction to G^0 due to the vector current in the limit $k'=0$

$$(\delta G^0)^V_{\text{total in the limit } k'=0} = \frac{3\alpha}{8\pi}G^0 D \tag{72}$$

which is *divergent* and independent of the details of the strong interaction. The usefulness of the above result will become clear shortly.

We now show that we can write

$$(\delta G^0)^V = \frac{3\alpha}{8\pi} G^0 D + \text{(soft photon portion).} \tag{73}$$

To see this [22, 23], let us first suppress the soft photons by introducing a low-energy cut-off λ on the photon propagator by writing

$$\frac{1}{Q^2} = \frac{1}{Q^2}\left(\frac{\lambda^2}{\lambda^2+Q^2}\right) + \frac{1}{\lambda^2+Q^2} \tag{74}$$

where $m_p^2 \gg \lambda^2 \gg m_e^2$. The first part dominates for low Q^2 and we can use in this case for the hadronic part of the amplitude (63) the Born terms only [24, 25] since these terms dominate for low Q^2. The second part of (74) dominates for large Q^2 and is less singular as $Q^2 \to 0$. In the corresponding part of the amplitude, one can neglect the lepton momenta whereupon one may use the result (72) with D given by

$$D = \ln\left(\frac{\Lambda^2}{\lambda^2}\right).$$

Thus we can write $(\delta G^0)^V$ in the form (73). On the other hand, the correction to G^0 from the electromagnetic corrections to the axial vector current (i.e. from the last term in (63)), namely $(\delta G^0)^A$ is, as we have seen, model-dependent and its divergent part is given by

$$G^0 \frac{3\alpha}{8\pi} 2\bar{Q}\left[\frac{1}{i\pi^2}\int \frac{d^4Q}{Q^4}\right].$$

If we cut off this integral at its lower end at the nucleon mass m_N and below that value use the Born approximation for $T^A_{\mu\nu}$ we can write the total correction as [23, 24]

$$(\delta G^0)^A = G^0 \frac{3\alpha}{8\pi} 2\bar{Q} \ln\left(\frac{\Lambda^2}{m_N^2}\right) + \text{axial Born terms.} \tag{75}$$

Thus the whole of the radiative correction to the vector coupling constant $G^{(0)}$ is given by (73) and (75):

$$\begin{aligned}\delta G^0 &= (\delta G^0)^V + (\delta G^0)^A \\ &= G^0 \frac{\alpha}{4\pi}\left\{\left[3\ln\left(\frac{\Lambda}{m_p}\right) + \text{(soft photon portion)}\right]\right. \\ &\quad \left. + 6\bar{Q}\ln\left(\frac{\Lambda}{m_p}\right) + \text{(small correction from axial Born terms)}\right\}\end{aligned} \tag{76}$$

Before we apply (76) to some particular cases, such as nuclear β-decay, the following remark is in order.

Our previous discussion depends on the use of the Bjorken limit as given by Eq. (9) and that the equal-time commutation relations appearing on the r.h.s. of (1) are the same as the "naive commutators" obtained by a straightforward use of canonical commutation relations in triplet models or in field algebra and equations of motion. An objection has been raised to the use of the Bjorken limit in the above sense by several people [25]. It has been shown that the Bjorken limit breaks down in second-order perturbation theory (of a renormalizable model of strong interactions) in the sense that for space-space commutators the Bjorken limit commutators and the "naive commutators" differ. However, for the time-time and time-space commutators, the Bjorken limit commutators and the naive commutators agree. This objection, if true, would affect only the model-dependent contribution $G^0\,(3\alpha/8\pi)2\bar{Q}D$ to the vector constant G^0 coming from the electromagnetic correction to the axial vector current. However, one should not take this objection too seriously, since there is no reason to believe that perturbation theory is relevant to the discussion of higher Q_0 behaviour of an amplitude. In any case, this objection would affect only the above-mentioned contribution which is model dependent anyway.

d) Applications

Nuclear β-Decay. We now apply Eq. (76) to nuclear β-decay. It has been shown that neglecting terms of order $\alpha E_e/m_p$, $\alpha E_e/m_p \ln m_p/E_e$ and $\alpha q/m_p$, q being the momentum transfer to the leptons, the soft photon portion in (73) or (76) is also independent of the details of the strong interactions, as well as of any current algebra. This soft photon portion of the second-order radiative corrections to the part of the electron spectrum arising from the vector current is given by a universal function [26]

$$\frac{\alpha}{2\pi} g(E_e, E_m, m_e) = \frac{\alpha}{2\pi}\left\{3\ln\frac{m_p}{m_e} - \frac{3}{4} + 4\left(\frac{\tanh^{-1}\beta}{\beta} - 1\right)\cdot\left[\frac{E_m - E_e}{3E_e} - \frac{3}{2} + \ln\frac{2}{m_e}(E_m - E_e)\right] + \frac{4}{\beta}L\left(\frac{2\beta}{1+\beta}\right) + \frac{\tanh^{-1}\beta}{\beta} \times\left[2(1+\beta^2) + \frac{(E_m - E_e)^2}{6E_e^2} - 4\tanh^{-1}\beta\right]\right\} \tag{77}$$

where p_e, E_e and E_m are the momentum, energy, end-point energy of the electron, respectively, $\beta = p_e/E_e$ and $L(x) = \int_0^x \frac{dt}{t} \ln(1-t)$ is the Spence function. The contribution of the soft photon portion to $(\delta G^0)^V$ is given by

$$\frac{\alpha}{4\pi}\left[3\ln\frac{m_p}{2E_m} - 5\right]. \tag{78}$$

Thus the second-order radiative corrections to the Fermi part of the electron spectrum can be written as [23] (p^0 denotes the spectrum in the absence of electromagnetic correction)

$$\begin{aligned} \Delta p\, d^3 p_e = p^0 d^3 p_e \Bigg\{ & \frac{\alpha}{2\pi}\left[3\ln\frac{\Lambda}{m_p} + g(E_e, E_m, m_e)\right. \\ & + \frac{\alpha}{2\pi}\left(6\bar{Q}\ln\frac{\Lambda}{m_p} - 1\right)\Bigg\} \end{aligned} \tag{79}$$

while

$$\begin{aligned} \delta G^0 &= (\delta G^0)^V + (\delta G^0)^A \\ &= G^0 \left\{ \frac{\alpha}{4\pi}\left[3\ln\frac{\Lambda}{m_p} + 3\ln\frac{m_p}{2E_m} - 5\right] \right. \\ &\quad \left. + \frac{\alpha}{4\pi}\left(6\bar{Q}\ln\frac{\Lambda}{m_p} - 1\right)\right\}. \end{aligned} \tag{80}$$

In (78) and (79) the expressions between the square brackets are, respectively, the model-independent contributions arising from the electromagnetic corrections to the vector current to the Fermi part of electron spectrum and to δG^0. They are therefore identical with the corresponding quantities in the "bare" particle calculation [27]. The last two terms in both (79) and (80) are the contributions arising from the electromagnetic corrections to the axial vector current, the last term in each case being the estimate of the small correction from the axial Born terms [23].

Collecting terms, we may finally write for the renormalized vector coupling constant G_V^β of nuclear β decay in the form

$$G_V^\beta = G^0 + \delta G^0 \tag{81}$$

where G^0 represents the "bare" (or original) coupling constant while G_V^β is the coupling constant directly measured. Hence

$$\begin{aligned} G^0 &= G_V^\beta - \delta G^0 \\ &= G_V^\beta\left[1-\frac{\delta G^0}{G_V^\beta}\right] = G_V^\beta\left[1-\frac{\delta G^0}{G^0}\right] \\ &= G_V^\beta\left[1-\frac{\alpha}{4\pi}\left(3(1+2\bar{Q})\ln\frac{\Lambda}{m_p}+3\ln\frac{m_p}{2E_m}-6\right)\right]. \end{aligned} \tag{82}$$

For comparison, we recall the corresponding result for the radiative correction to the μ-decay coupling constant which is finite and is given in (29b) and (30).

In order to apply formula (82) we must choose a beta emitter which is purely vector, and O^{14} is the best candidate for which

$$G_V^\beta = (1.4173 \pm 0.0034)\ 10^{-49}\ \text{ergs cm}^3\,. \tag{83}$$

Using now Eq. (82) and inserting for E_m the maximum positron energy 2.3 MeV from O^{14}, we list in Table 2 the values of G^0 deduced as a function of cut-off Λ for the "fractional" quark model ($\bar{Q}=1/6$) as well as the "integral" quark model ($\bar{Q}=-1/2$) for which no cut-off is needed.

Table 2. *Numerical estimates of G^0 and G^0/G as function of cut-off*

$\bar{Q}$	Cutoff in BeV	G^0 in 10^{-49} erg cm^3	G^0/G
1/6	300	1.3901	0.968 ± 0.002
	30	1.3975	0.973 ± 0.002
	10	1.4014	0.976 ± 0.002
	$\sim m_p$ (1 BeV)	1.4088	0.981 ± 0.002
$-1/2$	(no cutoff needed)	1.4088	0.981 ± 0.002

From Table 2 there appears to be a genuine ($\sim 2\%$) discrepancy between the "bare" leptonic coupling constant G and the "bare" semi-leptonic coupling constant G^0. In view of this difference, it is necessary to explore the possibility of departures from the universal coupling law in its simple form $G^0 = G$. The Cabibbo form of universality $G^0 = G\cos\theta$, $G^1 = G\sin\theta$, on the other hand, can not only explain the slight reduction of G^0 reletaive to G but it will account for the observed much larger reduction of G^1 (the "bare" $|\Delta Y|=1$ semi-leptonic coupling constant) with respect to G. One can now test the Cabibbo form of universality. From Table 2, the Cabibbo angle as determined from nuclear β-decay lies between the limits

$$0.968 < \cos\theta_V^\beta < 0.98$$

or

$$0.20 < \theta_V^\beta < 0.25 . \tag{84}$$

On the other hand, θ_V^M as determined from $K^+ \to \pi^0 e^+ \nu_e$ decay rate is given by [5]

$$\begin{aligned} f_+(0) \sin\theta_V^M &= (0.230 \pm 0.003),\ \lambda^+ = 0 \text{ (constant form factor)} \\ &= (0.224 \pm 0.003),\ \lambda^+ = 0.025 . \end{aligned} \tag{85a}$$

$f_+(0) = 1$, if one confines oneself to first-order $SU(3)$ symmetry breaking (Ademollo-Gatto theorem). Thus neglecting second-order $SU(3)$- breaking effects

$$\begin{aligned} \theta_V^M &= (0.230 \pm 0.003),\ \lambda^+ = 0 \\ &= (0.224 \pm 0.003),\ \lambda^+ = 0.025 . \end{aligned} \tag{85b}$$

θ_V^β and θ_V^M are consistent although θ_V^β is not fully unambiguously determined.

Remarks

i) As far as the radiative corrections to axial vector coupling constant G_A are concerned, one has the additional complication of the presence of

$$I_\lambda^M = \frac{e^2}{(2\pi)^4} \int \frac{d^4Q}{Q^2} \left[\frac{i}{2} \frac{\partial}{\partial q_\lambda} M_{\mu\mu}(p, p', Q, p)|_{q=p'-p} \right]$$

since the axial vector current is not conserved. In general, one cannot say anything about it or even whether it gives convergent or divergent contribution. The problem of the finiteness of the radiative corrections to the ratio $G_A/G^0 = G_A/G_V$ has been examined by various authors [28]. The models they considered are restrictive and involve some dynamical hypothesis, and for details the reader is referred to Ref. [28].

ii) If one gives up the local Fermi interaction and assumes weak decays to be mediated by charged vector bosons $W^\pm$ then it has been shown [29] that at *zero momentum transfer* to leptons the radiative corrections to the vector coupling constant come out to be divergent and independent of strong interactions. The divergent constant is independent of any model since the result depends only on the model-independent $SU(2)$ algebra of the time-time and time-space commutators. Moreover, the divergence comes out to be exactly the same as for μ-decay (note that the second-order radiative corrections to μ-decay are also divergent in this case). Thus as far as the question of comparison of G and G^0 is concerned for testing universality, the above situation may be adequate since a universal infinite constant can always be absorbed in the definition of the Fermi constant G. The final results [30], however, depend on the mass of W boson as well as on model-dependent constant $\bar{Q}$.

$\pi^{\pm} \to \pi^0 + e^{\pm} + \nu_e$

The radiative effects for the pion β-decay are most conveniently expressed as a correction to the form factor f_+, viz.

$$\delta f_+ = \frac{\alpha}{4\pi} f_+ \left[3\ln\frac{\Lambda}{m_p} + 3\ln\frac{m_p}{m_e} + 2(\xi + \ln(\eta x) - 1)\cdot\left(2\ln\frac{\lambda}{m_e} + 1\right) \right. \\ \left. - 2(\xi + \ln(\eta x))^2 + 4\ln(\eta x)\ln(1-\eta x) - 4L(\eta x) \right] + \frac{\alpha}{4\pi} f_+ 6\bar{Q}\ln\frac{\Lambda}{m_p} \tag{86}$$

where λ is the fictitious photon mass, $\xi = \ln m_\pi/m_e$, $\eta = 1 - (m_{\pi^0}/m_{\pi^+})^2$ and $x = p_e/p_e^{\max}$. In Eq. (86) the term within the square brackets is the model-independent contribution (also independent of the details of the strong interaction) arising from electromagnetic corrections to the vector current and is the same as in the "bare" particle case [24]. The second term of Eq. (86) is the model-dependent contribution arising from electromagnetic corrections to the axial current. If one integrates Eq. (86) over the electron spectrum, one finds

$$\frac{\Delta\tau}{\tau_0} = -\frac{\alpha}{2\pi}\left[3(1+2\bar{Q})\ln\frac{\Lambda}{m_p} + 3\ln\frac{m_p}{m_e} - 4\right] \\ - \frac{\alpha}{4\pi}\frac{30}{C_R}\left\{ -\frac{\xi}{15} - \frac{1}{15}\ln\eta - \frac{4\pi^2}{45} + \frac{257}{300} \right. \\ \left. + \eta\left[-\frac{1}{10}\xi - \frac{1}{6}\ln\eta - \frac{2\pi^2}{45} + \frac{1063}{1080}\right] + \ldots \right\} \tag{87}$$

where C_R is the recoil correction factor ~ 1. Hence, with a cut-off $\Lambda \cong m_p$ or for $\bar{Q} = -1/2$ (without cut-off), the fractional change in decay rate is $+1.2\%$. If we correct the theoretical prediction for the rate (based on the CVC hypothesis) by this amount, the value is

$$\tau^{-1} = (0.4048 \pm 0.04)\ \mathrm{sec}^{-1} \tag{88}$$

to be compared with the experimental value [31] $(0.393 \pm 0.027)\ \mathrm{sec}^{-1}$. The experimental accuracy is not sufficient as yet to deduce a value of the Cabibbo angle (as was done in connection with nucleon β-decay) but this may be possible in the future.

$\pi_{e2}/\pi_{\mu 2}$ ratio

For the radiative correction to the π_{e2} to $\pi_{\mu 2}$ ratio, we cannot apply the previous discussion due to the presence of unknown functions I_λ^M and I_λ^T in our general formula (52). However, from Eq. (62) it is easy to see that the divergent parts of the radiative corrections are the same for $\pi_{\mu 2}$ and π_{e2}. This implies that the radiative corrections to the ratio of these two rates is finite. In the absence of an exact calculation we give the perturbation-theoretic result. In giving this result, one must exercise some care because the usual experiment only counts electrons from π_{e2} whose energies are sufficiently high to avoid confusion with the electrons resulting from the sequence of decays $\pi \to \mu \to e$; this is not a problem with the muons from $\pi_{\mu 2}$. If ΔE_e is the energy interval below the maximum energy over which the π_{e2} electrons are accepted, the radiative correction will depend rather sensitively on ΔE_e. The following formula has been derived for the radiative correction to the $\pi_{e2}/\pi_{\mu 2}$ ratio [32]:

$$R = R_0 \left[0.861 + 0.0211 \ln \left(\frac{2 \Delta E_e}{m_e} \right) \right] \tag{89}$$

where the uncorrected ratio R_0 is

$$R_0 = \frac{m_e^2}{m_\mu^2} \left(\frac{m_\pi^2 - m_e^2}{m_\pi^2 - m_\mu^2} \right)^2 = 1.278 \times 10^{-4} . \tag{90}$$

In one experiment [33] which has been performed, $\Delta E_e = 10$ MeV so that the predicted value of R is 1.2×10^{-4}, to be compared with a measured value of $R = (1.18 \pm 0.07) \times 10^{-4}$. The most accurate experiment performed to date [34] measures the ratio $\dfrac{\Gamma(\pi \to e + \nu_e) + \Gamma(\pi \to e + \nu_e + \gamma)}{\Gamma(\pi \to \mu + \nu_\mu)}$ where the γ-ray comes from internal bremsstrahlung; the observed value is $(1.247 \pm 0.028) \times 10^{-4}$. The radiative corrections to this ratio have been calculated by *Kinoshita* [35] and their effect is to reduce R_0(Eq. (90)) by 3.9%. This reduction yields the value 1.228×10^{-4} in excellent agreement with the measurement. This extraordinary agreement is a great success for the (V, A) theory and implies equal coefficients for the electron and muon currents ("electron-muon universality") to better than 1%.

This completes our discussion of the electromagnetic radiative corrections to hypercharge-conserving semi-leptonic weak processes. The recent developments involving current algebra technique enabled us to calculate radiative corrections by taking account of strong interactions to all orders. The estimates are, in general, logarithmically divergent (except in certain models), but are sufficiently insensitive to the cut-off to justify some confidence in these estimates.

4. Radiative Corrections to $|\Delta Y|=1$ Semi-leptonic Processes

a) Divergences in Radiative Corrections to $|\Delta Y|=1$ Processes

The discussion of the divergences in the radiative corrections to $|\Delta Y|=1$ semi-leptonic processes using current algebra and the Bjorken technique is entirely parallel to that given in Section 3 for the $\Delta Y=0$ case. In fact, the expression for the divergent part of the radiative corrections to the hypercharge-changing processes $A\to B+\bar{l}+\bar{\nu}_e$ is immediately obtained from Eq. (62) with the replacements $G^0\to G^1$, $V_\lambda^0(x)\to V_\lambda^1(x)$, $A_\lambda^0(x)\to A_\lambda^1(x)$ and the constant c (Eq. (62a)) is replaced by another constant c^1, which is easily computed to be

$$c^1=-\frac{a_1}{2}+\frac{a_2}{6}-\frac{2a_3}{3} \tag{91}$$

where a_i are given in Table 1. Now, from the conditions that the $\Delta Y=0$ and $|\Delta Y|=1$ currents J_λ^{1+i2} and J_λ^{4+i5}, respectively, should have charge equal to $+1$, we find that

$$a_1=a_2=1 \tag{91a}$$

and hence

$$c^1=-\frac{1}{3}-\frac{2a_3}{3}. \tag{92}$$

In models where the $|\Delta Y|=1$ current is not the $(4+i5)$ but some other component of an octet, the constant c^1 will be different but it can be easily computed.

With these preliminaries we may write for the divergent part of the radiative corrections to the $|\Delta Y|=1$ process $A\to B+l^-+\bar{\nu}_l$ the expression (see Eq. (62))

$$\begin{aligned}(\delta T_0)_{\text{div}}&=(T_L+T_H+T_{HL})_{\text{div}}\\ &=\left(-\frac{ie^2}{(2\pi)^4}\int\frac{\mathrm{d}^4Q}{Q^4}\right)\frac{3}{2}(1-c^1)T_0+\frac{G^1}{\sqrt{2}}(I_\lambda^M+I_\lambda^T)_{\text{div}}\,l_\lambda\end{aligned} \tag{93}$$

where T_0 is the uncorrected matrix element and I_λ^M and I_λ^T are the quantities defined previously with $J_\lambda^0\to J_\lambda^1$. From this expression we see that cancellation of divergences is quite a different problem in the present case from what it was in the $\Delta Y=0$ case. There, for the vector current taken between two members of the same isomultiplet, the two quantities I_λ^M and I_λ^T could be shown to be of higher order in e and so neglected. Instead, the problem here is similar to that of the axial vector current in the hypercharge-conserving case; that is to say, the quantity I_λ^M cannot be neglected since the divergence neither of the vector nor of the axial vector part of $J_\lambda^I(x)$ vanishes to zero order in e. With regard to the term I_λ^T, we recall that this quantity could be neglected in the hypercharge-conserving case for transitions between members of the same isomultiplet because its proportionality to q, the momentum transfer, made it of order α^2 in that case. For hypercharge-changing transitions, however, this is no longer true and one is obliged to retain this term in general. One may, of course, calculate in the limit $q=0$ where upon the term I_λ^T vanishes; but this is a rather unrealistic limit for hypercharge-changing processes which can involve quite large momentum transfer. Thus the problem of dealing with I_λ^M and I_λ^T cannot be circumvented in the hypercharge-changing case.

Let us for the moment set aside this problem and simply require that the divergences which we can see in Eq. (93) should cancel; we find that we must have

$$c^1 = 1 . \tag{94}$$

This condition, together with (92) yields

$$a_3 = -2 . \tag{94a}$$

By inspection of Eqs. (66b), one sees immediately that Eq. (94a) implies that the third member of the underlying fundamental triplet (if there be such) possesses $Y=-2$ and therfore $Q=-1$. Thus, for $(V-A)$ currents, which are the $(1+i2)$ and $(4+i5)$ components of an octet in the $\Delta Y=0$ and $|\Delta Y|=1$ cases, respectively, the cancellation of the obvious divergences in the radiative corrections (there may, of course, also be divergences in the term $I_\lambda^M + I_\lambda^T$ in Eq. (93) but we have no information concerning this at present) requires that the basic triplet possess the quantum numbers of the particles Ξ^0, Ξ^- and Ω^-. If we change our current to the $(V+A)$ form, a similar argument to the above leads to the conclusion that a fundamental triplet of the type (p, n, Λ) will meet the cancellation (of divergences) requirement.

We shall not proceed further with the present treatment of radiative corrections to $|\Delta Y|=1$ processes because of the difficulties connected with the presence of the quantities I_λ^M and I_λ^T. Instead, we shall turn to ordinary perturbation theory to estimate radiative corrections to processes of interest.

b) Radiative Corrections to K_{l3} Decay

Radiative corrections to K_{l3} decay have been computed using straightforward perturbation theory [36]. These corrections are of interest for at least two reasons. In the first place, an accurate determination of the quantity $f_+(0)\sin\theta$ from the $K_{e3}^\pm$ rate requires a knowledge of the radiative corrections to this rate. Secondly, the value of the parameter $\xi(0)=f_-(0)/f_+(0)$ obtained from the ratio of the $K_{\mu 3}^\pm$ and $K_{e3}^\pm$ rates is, in certain circumstances sensitive to the exact value of this ratio.

The results of the perturbation-theoretic calculations which, as usual, contain a logarithmic divergence, are as follows:

1) The radiative corrections to the Dalitz plot in $K^{\pm}_{e3}$ and K^{0}_{e3} is substantial and averages more than 3% in absolute magnitude over a good portion of this plot. This result is quite insensitive to values of the cut-off (Λ) varying between one and two nucleon masses (m_N).

2) The radiative corrections to the electron spectra in $K^{\pm}_{e3}$ and K^{0}_{e3} are noticeably different, being everywhere more positive in the former case. This result is also relatively insensitive to the value of Λ.

3) With regard to the rates $\Gamma(K^{\pm}_{e3})$ and $\Gamma(K^{0}_{e3})$, the results are logarithmically dependent on Λ. For $\Lambda \cong m_N$, the fractional changes in these quantities are

$$\frac{\Delta\Gamma(K^{\pm}_{e3})}{\Gamma(K^{\pm}_{e3})} = 0.8\% \tag{95a}$$

and

$$\frac{\Delta\Gamma(K^{0}_{e3})}{\Gamma(K^{0}_{e3})} = -0.45\% \tag{95b}$$

respectively. The radiative correction to the ratio of the rates is independent of Λ and is 1.25%.

4) For $K_{\mu 3}$ decays, the radiative corrections to the Dalitz plot are not more than a fraction of a per cent throughout the whole energy range for values of $\xi(0)$ ranging from -1 to 1.

5) The radiative corrections to the $K_{\mu 3}$ rate are sensitive both to Λ and to the value of $\xi(0)$. For a typical value of $\xi(0) \cong 0.1$ and $\Lambda \cong m_N$, it is estimated that

$$\frac{\Delta\Gamma(K_{\mu 3})}{\Gamma(K_{\mu 3})} \cong 0.07\%\,. \tag{96}$$

6) From the numbers given in Eqs. (95a) and (96), we estimate the correction to the ratio $\Gamma(K^{\pm}_{\mu 3})/\Gamma(K^{\pm}_{e3})$ as about -0.7%. This result is independent of the cut-off and it indicates that the radiative corrections to this ratio are quite small. This statement is subject to the qualification that the axial vector current contribution (which contributes to the vector coupling constant and is model-dependent, as we have seen in the current algebra treatment of $\pi^{\pm} \to \pi^{0} + e^{\pm} + \nu_e$ (Section 3) has not been taken into account.

c) Radiative Corrections to $K^{\pm}_{e2}/K^{\pm}_{\mu 2}$ Ratio

The ratio of the $K^{\pm}_{e2}$ and $K^{\pm}_{\mu 2}$ rates is a very important test of the (V, A) assumption regarding the nature of hadron current in hypercharge-changing weak interactions. It is therefore also important to have an estimate of the radiative corrections to this ratio. Such an estimate is similar to the one for the $(\pi_{e2}/\pi_{\mu 2})$ ratio and the result is [37]

$$R = R_0 \left[0.815 + 0.0273 \ln \left(\frac{2\Delta E}{m_e} \right) \right] \tag{97}$$

where R_0 is the uncorrected ratio

$$R_0 = \left(\frac{m_e}{m_\mu} \right)^2 \left(\frac{m_K^2 - m_e^2}{m_K^2 - m_\mu^2} \right)^2 = 2.58 \times 10^{-5} \tag{98}$$

and ΔE is the energy interval below $(E_e)_{\max} = (m_K^2 + m_\pi^2)/2m_K$ in which electrons are accepted as arising from K_{e2} in the course of an actual experiment. The discussion of the experimental results applying Eq. (97) are given in Ref. [38]. The constant term (0.815) on the r.h.s. of Eq. (97) represents the suppression of the uncorrected ratio due to the virtual photons. The term depending on the energy resolution ΔE takes account of the inner bremsstrahlung within this energy interval. The most accurate published experiment [38], allowing for the inner bremsstrahlung effect, yields $R = 1.9 {+0.7 \atop -0.5} \times 10^{-5}$; this is therefore to be compared with the theoretical prediction $0.815\, R_0 = 210 \times 10^{-5}$. The agreement is excellent and constitutes a strong argument for the dominance of the axial vector current in this decay.

Acknowledgments. The author is very grateful to Dr. *C. Ryan* for reading through the manuscript. He would also like to thank Professor's *Abdus Salam* and *P. Budini* and the International Atomic Energy Agency for hospitality at the International Centre for Theoretical Physics, Trieste.

References and Footnotes

1. See, for example, *Adler, S. L., Dashen, R. F.*: Current algebras. New York: W. A. Benjamin, Inc. 1968.
2. See, for example, *Okubo, S.*: Nuovo Cimento **44**A, 1015 (1966).
3. See, for example, *Wu, C. S.*: Rev. Mod. Phys. **36**, 618 (1964).
4. See, for example, *Schopper, H, F.*: Weak interactions and nuclear β-decay. Amsterdam: North-Holland 1965. The value of G_V^β quoted on p. 52 of this reference is with radiative corrections. The value quoted in Eq. (6) of these lectures is without radiative corrections.
5. See, for example, *Rubbia, C.*: Topical Conference on Weak Interactions, CERN, Geneva, January 1969.
6. *Bjorken, J. D.*: Phys. Rev. **148**, 1467 (1966).
7. *Berman, S. M.*: Phys. Rev. **112**, 267 (1958).
8. *Kinoshita, T., Sirlin, A.*: Phys. Rev. **113**, 1652 (1958).
9. See, for example, *Jauch, J. M., Rohrlich, F.*: The theory of photons and electrons. p. 185—187. Reading: Addison-Wesley 1959. Our definition of $\Sigma^{(2)}(p)$ is such that it differs from $\Sigma^{(2)}(p)$ of this reference by a minus sign.
10. *Berman, S. M., Sirlin, A.*: Ann. Phys. (N.Y.) **20**, 20 (1962).
11. If Seagull terms are present we shall not write them explicitly. The * on T-product indicates their possible presence.

12. Techniques of current algebra in the study of radiative corrections to weak interactions were first applied by *Bjorken, J. D.*, Ref. 6, and Phys. Rev. **160**, 1582 (E) (1967) and *Abers, E. S., Norton, R. E., Dicus, D. A.:* Phys. Rev. Letters **18**, 676 (1967). These authors did not consider the contribution coming from the electromagnetic corrections to the axial vector current. This was considered in Ref. 13.
13. *Cabibbo, N., Maiani, L., Preparata, G.:* Phys. Letters **25**B, 31, 1387 (1967). — *Johnson, K., Low, F. E., Surra, H.:* Phys. Letters **18**, 1224 (1967).
14. One does not know whether or not there are operator Schwinger terms in equal-time commutation relations. If they are present, we are assuming that, in the divergence relations like (44) or (39), they are cancelled by the Seagull terms that might be present in our matrix elements (recall that * on *T*-product was put to indicate the possible presence of such terms). See, for example, *Brown, S. C.:* Phys. Rev. **158**, 1444 (1967); **163**, 1854 (E) (1967).
15. *Nakamura, K.:* Tokyo University preprint (1968). — *Weisberger, W., Preparata, G.:* Phys. Rev. **175**, 1965 (1968). — *Sirlin, A.:* Phys. Rev. **176**, 1871 (1968).
16. *Lee, T. D., Weinberg, S., Zumino, B.:* Phys. Rev. Letters **18**, 1029 (1967).
17. *Okubo, S.:* Ann. Phys. (N.Y.) **38**, 337 (1966); cf. also *Marshak, R. E.:* Proc. Intern. Spring School at Yalta (1966), p. 139.
18. *Gell-Mann, M.:* Phys. Letters **8**, 214 (1964). — *Zweig, G.:* CERN preprint 8182/Th. 401 (1964).
19. *Sakata, S.:* Progr. Theoret. Phys. (Kyoto) **16**, 686 (1956).
20. *Maki, Z.:* Progr. Theoret. Phys. (Kyoto) **31**, 331 (1964). — *Hara, Y.:* Phys. Rev. **134**, B 701 (1964).
21. *Abers, E. S., Norton, R. E., Dicus, D. A.:* Ref. 12.
22. *Bjorken, J. D.:* Lectures given at the Intern. School of Physics, Enrico Fermi Varenna, 1967.
23. *Abers, E. S., Dicus, D. A., Norton, R. E., Quinn, H. R.:* Phys. Rev. **167**, 1461 (1968).
24. See Refs. 8 and 10. For $\pi\beta$-decay, see *Chang, N.:* Phys. Rev. **131**, 1272 (1963).
25. *Adler, S., Tung, Wu-Ki:* Phys. Rev. Letters **22**, 978 (1969); *Jackiw, R., Preparata, G.:* Phys. Rev. Letters **22**, 975, 1162 (E) (1969).
26. *Sirlin, A.:* Phys. Rev. **164**, 1767 (1967).
27. *Berman, S., Sirlin, A.:* Ref. 10. All that we have to do is to set $\varrho = 0$ and neglect $C_{1,2} \to 0$ in their calculation. Cut-off Λ is now, of course, independent of nucleon structure.
28. *Sirlin, A.:* Ref. 15. — *Weisberger, W., Preparata, G.:* Ref. 15. — *Callan, C. G.:* Phys. Rev. **169**, 1175 (1968). — *de Alwis, S. P.:* Cambridge University preprint (1968).
29. — Phys. Rev. Letters **19**, 877 (1967).
30. — Proc. 14th Intern. Conference on High-Energy Physics, Vienna (1968), p. 321.
31. *Rosenfeld, A. H., et al.:* Rev. Mod. Phys. **41**, 109 (1969).
32. *Berman, S.:* Phys. Rev. Letters **1**, 468 (1968).
33. *Anderson, C., et al.:* Phys. Rev. **119**, 2050 (1960).
34. *Di Capua, E., et al.:* Phys. Rev. **133** B, 1333 (1964).
35. *Kinoshita, T.:* Phys. Rev. Letters **2**, 477 (1959).
36. *Ginsberg, E. S.:* Phys. Rev. **142**, 1035 (1966); **162**, 1570 (1967).
37. *Berman, S.:* Ref. 32. — *Neville, D. E.:* Phys. Rev. **124**, 2037 (1960); see also *Chang, N. P.:* Phys. Rev. **129**, 399 (1963).
38. *Botterill, D. R., et al.:* Phys. Rev. Letters **19**, 982 (1967).

Dr. *Riazuddin*
Institute of Physics, University of Islamabad
Rawalpindi, Pakistan
and
International Centre for Theoretical Physics
Trieste, Italy

Radiative Corrections to Weak Interactions

J. ROTHLEITNER

Contents

I. Introduction 161
II. Radiative Corrections to Leptonic Processes 163
III. Radiative Corrections to Semileptonic Decays 168
References 170

I. Introduction

Disregarding the effects of CP-violation (of which we even don't know, whether they originate in the weak interaction) we seem to have achieved a satisfactory formulation of the weak interaction in the form of the universally self-coupled weak current:

$$\mathscr{L}_{\mathrm{w}}(x) = \frac{G}{\sqrt{2}} \lim_{\xi=0} : \frac{1}{4}\left\{ J_\lambda\left(x+\frac{\xi}{2}\right) J^{+\lambda}\left(x-\frac{\xi}{2}\right) + J_\lambda\left(x-\frac{\xi}{2}\right) J^{+\lambda}\left(x+\frac{\xi}{2}\right) \right.$$
$$\left. + J_\lambda^+\left(x+\frac{\xi}{2}\right) J^\lambda\left(x-\frac{\xi}{2}\right) + J_\lambda^+\left(x-\frac{\xi}{2}\right) J^\lambda\left(x+\frac{\xi}{2}\right)\right\}: \tag{1}$$

$\mathscr{L}_{\mathrm{w}}$ is hermitean and (in the limit $\xi=0$) gauge invariant.

The weak current $J_\lambda(x)$ is characterized by the properties:

1) $J_\lambda(x)$ is left-chiral i.e. $V-A$.

2) $J_\lambda(x)$ raises the charge by one unit:

$$[Q, J_\lambda(x)] = J_\lambda(x)\,. \tag{2}$$

3) $J_\lambda(x)$ is the sum of a hadronic and a leptonic piece:

$$J_\lambda = h_\lambda + l_\lambda \tag{3}$$

and is normalized uniquely (apart from a phase factor, which is irrelevant in $\mathscr{L}_{\mathrm{w}}$) by the equal time commutation relation:

$$[J_0^3(\boldsymbol{x}, t), J_0(\boldsymbol{y}, t)] = 4\,\delta^3(\boldsymbol{x}-\boldsymbol{y})\,J_0(\boldsymbol{x}, t) \tag{4}$$

where J_0^3 is defined by

$$[J_0(\boldsymbol{x}, t), J_0^+(\boldsymbol{y}, t)] =: \delta^3(\boldsymbol{x}-\boldsymbol{y})\,2J_0^3(\boldsymbol{x}, t)\,. \tag{5}$$

If we assume that the leptonic piece is to be constructed from the muonic and electronic fields $\Psi_\mu(x)$, $\Psi_{\nu_\mu}(x)$, $\Psi_e(x)$ and $\Psi_{\nu_e}(x)$, l_λ is determined by 1), 2), 3):

$$l_\lambda = e^{i\alpha}\,\overline{\psi}_{\nu_\mu}\gamma_\lambda(1-\gamma_5)\,\psi_\mu + e^{i\beta}\,\overline{\psi}_{\nu_e}\gamma_\lambda(1-\gamma_5)\,\psi_e \tag{6}$$

where α, β are real, but otherwise arbitrary parameters.

If we assume that the hadronic piece is to be constructed from the 8 hadronic vector currents (CVC) and the 8 hadronic axialvector currents (normalized by the equal time commutators of the 0-components as given by *Gell-Mann* [1]), h_λ is not uniquely determined by 1), 2), 3), it contains 2 arbitrary (angle-)parameters:

$$h_\lambda = \cos\theta\, h_\lambda^{\pi^+} + \sin\theta\, e^{i\gamma}\, h_\lambda^{K^+}\,, \tag{7}$$

θ = *Cabibbo* [2] angle, ($\theta \approx 0.23$ from fitting the data [3]).

If electron- and muon-number are strictly conserved (as we believe so far) α and β are unmeasurable, the respective phase factors in (6) can be absorbed by a redefinition of the lepton fields. Analogously the phase factor $e^{i\gamma}$ in (7) can be absorbed by redefinition of $h_\lambda^{K^+}$, provided the other interactions allow this.

$\mathscr{L}_w$ may be decomposed into a nonleptonic, a semileptonic and a leptonic part:

$$\mathscr{L}_w = \frac{G}{\sqrt{2}}\, h_\lambda^+ h^\lambda + \frac{G}{\sqrt{2}}\,(h_\lambda^+ l^\lambda + l_\lambda^+ h^\lambda) + \frac{G}{\sqrt{2}}\, l_\lambda^+ l^\lambda\,. \tag{8}$$

The problems connected with this form of $\mathscr{L}_w$ are:

a) It is a nonrenormalizeable interaction and so far nobody is able to calculate convincingly second-or higher-order amplitudes. However, since the interaction is weak for low energies one can hopefully restrict one-self to first order perturbation theory for low energies. At energies $\gtrsim 300$ GeV this certainly is not possible, as the first order amplitudes violate unitarity for these energies. Restricting to first order perturbation theory $\mathscr{L}_w$ gives quite a good description of the leptonic and semileptonic weak interaction phenomena as we heard in the lectures by Prof. *Gasiorowicz*.

b) The situation is not so clear for the nonleptonic part of $\mathscr{L}_w$. This is, however, due to our poor understanding of strong interactions. Current algebra has brought much progress here, but it has also difficulties, as is discussed in Prof. *Stech*'s lectures.

c) Neglecting electromagnetic effects the matrix elements for the semileptonic processes are largely determined by the symmetries of the strong interactions, the purely leptonic processes are completely calculable to order G. The values for the Fermi-coupling constant G obtained in this way from μ-decay and e.g. O^{14} β-decay agree within 2%, indicating that there is a universal coupling. We should, however, be able to calculate the relevant corrections which have their origin in the electromagnetic interaction. These calculations are plagued with infinities [4–8] which lead various authors to various conclusions.

The problem of the "divergence" of radiative corrections to semi-leptonic weak interactions is discussed in Prof. *Riazuddin*'s lectures; in my lecture a discussion of the finite radiative correction to low energy β-decay will be given, which are of practical importance. Also a general discussion of the radiative corrections to purely leptonic weak interactions will be given. Here pure QED is involved and we can hope to obtain insight into the fundamental question whether the "divergences" have their origin in an inconsistent form of the weak Lagrangian or in our inability to treat QED in the correct way.

II. Radiative Corrections to Leptonic Processes

Since we restricted to first order in the weak interaction the problem is a purely electrodynamic one. Quantum electrodynamics is the most developed theory we have, and we should accordingly largely be able to solve the problem. This is, however, still not the case. Let us start with

μ-Decay

The Fierz reordering theorem allows us to rewrite the relevant part of $\mathscr{L}_w$ in the charge retaining form:

$$\mathscr{L}_w = \frac{G}{\sqrt{2}} \overline{\psi}_e \gamma_\lambda (1-\gamma_5)\psi_\mu \overline{\psi}_{\nu_\mu} \gamma^\lambda (1-\gamma_5)\psi_\nu + \cdots . \tag{9}$$

This enormously simplifies the problem, since we have only to consider the vertexfunctions:

$$\begin{aligned} -\int d^4x\, d^4y\, e^{ip'x} e^{-ipy} \langle 0|\mathscr{T}\, \psi_e(x)\, \overline{\psi}_\mu(y)\, v_\lambda^{e\mu}(0)|0\rangle \\ =: S'_{Fe}(p')\, \Gamma_\lambda^{e\mu}(p',p)\, S'_{F\mu}(p), \end{aligned} \tag{10}$$

$$\begin{aligned} -\int d^4x\, d^4y\, e^{ip'x} e^{-ipy} \langle 0|\mathscr{T}\, \psi_e(x)\, \overline{\psi}_\mu(y)\, a_\lambda^{e\mu}(0)|0\rangle \\ =: S'_{Fe}(p')\, \Gamma_{\lambda 5}^{e\mu}(p',p)\, S'_{F\mu}(p) \end{aligned} \tag{11}$$

where we defined

$$v_\lambda^{e\mu} = \overline{\psi}_e \gamma_\lambda \psi_\mu, \qquad a_\lambda^{e\mu} := \overline{\psi}_e \gamma_\lambda \gamma_5 \psi_\mu . \tag{12}$$

It is gratifying that one finds:

1) The corrections are finite to every order in α [4].

2) The calculated corrections to order α which come from the graphs

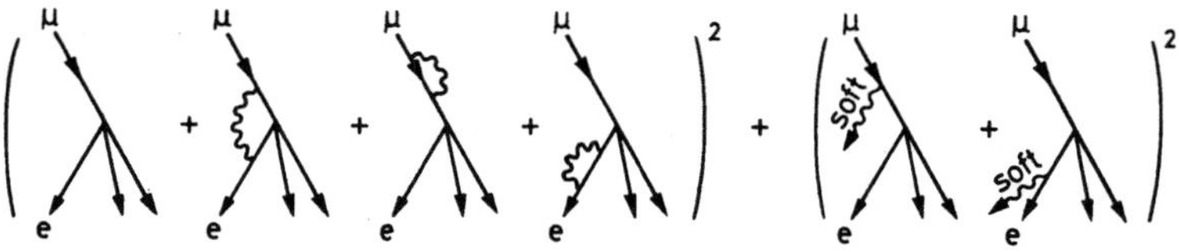

are definitely confirmed by experiment. According to the calculations by *Kinoshita* and *Sirlin* [9] the decay spectrum receives corrections of 5% to 6% in the region $0.3 \leqq \frac{Pe}{Pe\,\mathrm{max}} \leqq 0.95$ and the angular distribution about 3%. At the upper end of the spectrum the corrections are even larger*.

Using the corrected amplitudes recent values for the Michel parameter ϱ and the angular distribution parameter δ are

$$\begin{aligned} \varrho &= 0.747 \pm 0.005 \text{ [10]} \\ \delta &= 0.758 \pm 0.010 \text{ [11]} \end{aligned} \qquad [\varrho = \delta = \tfrac{3}{4} \text{ in } V-A \text{ theory}].$$

The correction to the lifetime is given by

$$\frac{1}{\tau} = \frac{1}{\tau_0}[1\text{–}0.042], \qquad \text{from which one obtains}$$

$$G = (1.43506 \pm 0.00026) \times 10^{-49} \text{ erg cm}^3 =: G_\mu .$$

Let us discuss now the proof of 1):

Denoting $\partial_\lambda v^\lambda =: d$, $\partial_\lambda a^\lambda =: d_5$ we can derive the Ward-like identities ($\alpha, \beta = e, \mu$)

$$(p'-p)^\lambda \Gamma_\lambda^{\alpha\beta}(p', p) = -[S'^{-1}_{F\alpha}(p') - S'^{-1}_{F\beta}(p)] + \Gamma_d^{\alpha\beta}(p', p), \tag{13}$$

$$(p'-p)^\lambda \Gamma_{\lambda 5}^{\alpha\beta}(p', p) = -[S'^{-1}_{F\alpha}(p')\gamma_5 + \gamma_5 S'^{-1}_{F\beta}(p)] + \Gamma_{d_5}^{\alpha\beta}(p', p). \tag{14}$$

We know that $S'_{F\alpha}(p)$ is multiplicatively renormalizeable i.e.

$$\tilde{S}'_{F\alpha}(p) = \frac{1}{Z_\alpha} S'_{F\alpha}(p) \qquad \text{has a power series expansion** in } \alpha.$$

Assuming, that also all the vertexfunctions Γ are multiplicatively renormalizeable, i.e. that

$$\tilde{\Gamma}_A(p', p) = Z_A \Gamma_A(p', p) \qquad \text{have a power series expansion**},$$

* *Pe* and *Pe* max denote the momentum of the electron and its maximum value respectively.

** By this we mean, that in renormalized perturbation theory, any finite order gives a finite result. It is not claimed, however, that the perturbation series converges.

we can write

$$(p'-p)^{\lambda}\tilde{\Gamma}_{\lambda}^{\alpha\beta}(p',p) = -\left[\frac{Z_v}{Z_\alpha}\tilde{S}_{\mathrm{F}\alpha}'^{-1}(p') - \frac{Z_v}{Z_\beta}\tilde{S}_{\mathrm{F}\beta}'^{-1}(p)\right] + \frac{Z_v}{Z_d}\tilde{\Gamma}_d^{\alpha\beta}(p',p) \tag{15}$$

and a similar equation for $\Gamma_{\lambda 5}$.

We remark that if one has to use a cut off Λ in perturbation theory (for whatever reason) the $\tilde{\Gamma}$ and $\tilde{S}$ are cut off dependent, they however have a finite and well defined limit as $\Lambda \to \infty$*. Taking Eq. (15) between spinors $\bar{u}(p')$ and $u(p)$ and putting $p'^2 = \breve{m}_\alpha^2$, $p^2 = \breve{m}_\beta^2$ we get

$$\bar{u}(p')(p'-p)^{\lambda}\tilde{\Gamma}_{\lambda}^{\alpha\beta}(p',p)\,u(p)\Big|_{\substack{p'^2=\breve{m}_\alpha^2\\ p^2=\breve{m}_\beta^2}} = \frac{Z_v}{Z_d}\bar{u}(p')\tilde{\Gamma}_d^{\alpha\beta}(p',p)\,u(p)\Big|_{\substack{p'^2=\breve{m}_\alpha^2\\ p^2=\breve{m}_\beta^2}} \tag{16}$$

letting $\Lambda \to \infty$, the left hand side has a finite limit, so the right hand side must also have a finite limit, i.e. $\lim\limits_{\Lambda=\infty}\dfrac{Z_v}{Z_d}$ = finite.

Multiplying Eq. (15) now from the left by $\bar{u}(p')$ and putting $p'^2 = \breve{m}_\alpha^2$ we conclude in the same way: $\lim\limits_{\Lambda=\infty}\dfrac{Z_v}{Z_\beta}$ = finite, and finally also: $\lim\limits_{\Lambda=\infty}\dfrac{Z_v}{Z_\alpha}$ = finite.

In completely the same manner we proceed with the axial vertex-function and conclude that also

$$\frac{Z_a}{Z_{d_5}},\quad \frac{Z_a}{Z_\beta}\quad \text{and}\quad \frac{Z_a}{Z_\alpha}\quad \text{have finite limits for}\quad \Lambda = \infty\,.$$

The Matrixelements we are interested in:

$$\langle e|v_\lambda^{e\mu}|\mu\rangle = \sqrt{\frac{Z_e}{Z_v}}\sqrt{\frac{Z_\mu}{Z_v}}\,\bar{u}(e)\,\tilde{\Gamma}_\lambda^{e\mu}\,u(\mu)$$

and (17)

$$\langle e|a_\lambda^{e\mu}|\mu\rangle = \sqrt{\frac{Z_e}{Z_a}}\sqrt{\frac{Z_\mu}{Z_a}}\,\bar{u}(e)\,\tilde{\Gamma}_{\lambda 5}^{e\mu}\,u(\mu)$$

are therefore finite!

Now we have to inquire under what conditions the multiplicative renormalizeability of the various vertexfunctions holds:

$$\Gamma_\lambda^{\alpha\beta}(p',p) \quad \text{and} \quad \Gamma_{\lambda 5}^{\alpha\beta}(p',p)$$

belong to bilinear spinor products therefore they are determined by the linear integral equation

$$\Gamma_{\lambda(5)}^{\alpha\beta}(p',p) = \gamma_\lambda(\gamma_5) - \int \Gamma_{\lambda(5)}^{\alpha\beta}\, S_{\mathrm{F}\alpha}' S_{\mathrm{F}\beta}' K^{\alpha\beta}\,. \tag{18}$$

* This makes the "proof" of *Preparata* and *Weisberger* [7] incomplete.

General renormalization theory [12] tells us that these Γ's are multiplicatively renormalizeable, *provided* that part of the scattering kernel $K^{\alpha\beta}$, which can be decomposed into

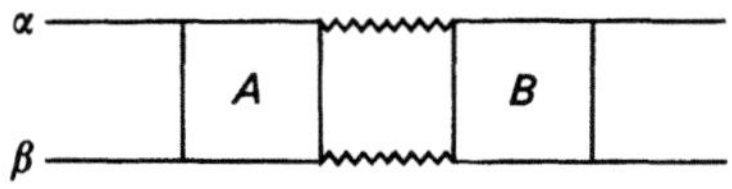

does not contribute to the integral above. If $\alpha = \mathrm{e}$, $\beta = \mu$ this certainly holds because of μ-conservation.

In order that the same argument applies also to Γ_d and Γ_{d_5}, they also must be bilinear spinor products, i.e. we must have the "partial" conservation laws

$$\partial^\lambda v_\lambda^{\alpha\beta} = m^{\alpha\beta}\, \overline{\psi}_\alpha \psi_\beta \quad \text{and} \quad \partial^\lambda a_\lambda^{\alpha\beta} = m_5^{\alpha\beta}\, i\, \overline{\psi}_\alpha \gamma_5 \psi_\beta \tag{19}$$

where $m^{\alpha\beta}$, $m_5^{\alpha\beta}$ are constants.

For $\alpha \neq \beta$ this is true in perturbation theory, hence the finiteness to any order of the corrections for μ-decay and $\nu_\mu + \mathrm{e} \to \mu + \nu_\mathrm{e}$ follows.

Elastic ν_l l-Scattering

This process is (after Fierz reordering) contained in

$$\mathscr{L}_\mathrm{w} = \frac{G}{\sqrt{2}}\, \bar{l}\gamma_\lambda(1-\gamma_5)\, l \cdot \bar{\nu}_l \gamma^\lambda (1-\gamma_5)\, \nu_l + \cdots . \tag{20}$$

Again we have to study only the vertexfunctions, this time however the diagonal ones: $\alpha = \beta = l$.

Here the dangerous part of the scattering kernel

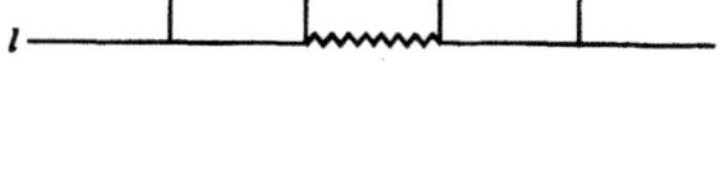

does contribute in $\Gamma_{\lambda 5}^{ll}$ (it does not contribute to Γ_λ^{ll} because of charge conjugation invariance), hence $\Gamma_{\lambda 5}^{ll}$ is *not necessarily* multiplicatively renormalizeable. But, even if it was we would also need the multiplicative renormalizeability of the divergence of the axial current in order to be able to conclude that $\dfrac{Z_l}{Z_a}$ is finite. That the divergence is *not* multiplicatively renormalizeable has been shown by *Adler** [8]. Hence the proof of the finiteness of $\frac{Z_l}{Z_a}$ breaks down and *Adler* indeed finds that $Z_l \bar{u}(p')\,\Gamma_{\lambda 5}(p', p)\,u(p)$ does not have a finite limit as $\Lambda \to \infty$. The trouble shows up for the first time to order α^2 and clearly comes from the graph

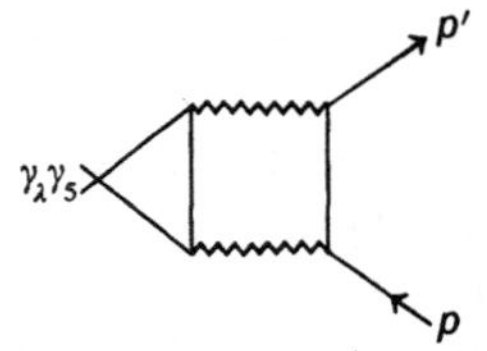

Adler finds

$$Z_l \bar{u}(p')\,\Gamma_{\lambda 5}(p', p)\,u(p) = \left[1 - \frac{3}{4}\left(\frac{\alpha}{\pi}\right)^2 \ln \frac{\Lambda^2}{m^2}\right] \bar{u}(p')\,\gamma_\lambda \gamma_5\,u(p) + \text{finite}. \quad (21)$$

Thus the radiative corrections are definitely not finite in a power series expansion in α.

Three points of view can be taken towards this result:

1) The local current-current theory of leptonic weak interactions cannot be correct, even if taken in lowest order. Despite its impressive success in describing the μ-decay we have to modify it, hoping that the modifications, while preserving the success in μ-decay, will at the same time cure the disease of not allowing a power series expansion of the radiative corrections.

2) Maintain the local current-current form to lowest order, but modify the Langrangian by adding $\nu_e\mu$ and $\nu_\mu e$ scattering terms, to wit:

$$\mathscr{L}_w = \frac{G}{\sqrt{2}} \{[\bar{\psi}_\mu \gamma_\lambda (1-\gamma_5)\,\psi_\mu - \bar{\psi}_e \gamma_\lambda (1-\gamma_5)\,\psi_e]\,[\bar{\psi}_{\nu_\mu} \gamma^\lambda (1-\gamma_5)\,\psi_{\nu_\mu} - \bar{\psi}_{\nu e} \gamma^\lambda (1-\gamma_5)\,\psi_{\nu e}]\} + \cdots. \quad (22)$$

Then the electron and muon triangle graphs cancel each other. The new Lagrangian can be tested by looking for elastic scattering of muon neutrinos by electrons (*Adler*).

---

* *Adler* traces this back to his claim, that the diagonal axialvector current is not "partially" conserved, instead he claims the relation

$$\partial_\lambda a^\lambda = 2m_0\,\bar{\psi} i \gamma_5\,\psi + \frac{\alpha_0}{4\pi}\,\varepsilon_{\mu\nu\varrho\sigma} F^{\mu\nu} F^{\varrho\sigma}.$$

This was discussed also in the lectures by Prof. *Wess* and Prof. *Zumino*. The essential features are already contained in a paper of *J. Schwinger*, P. R. **82**, 664 (1951).

3) Why should we expect, that radiative corrections can be expanded in a power series in α? If this is impossible, the infinity merely reflects the fact, that we are trying to do what is impossible. This would solve all our "radiative troubles": The possibility of a power expansion of the radiative corrections to μ-decay is then a lucky accident*.

III. Radiative Corrections to Semileptonic Decays

It is well known and is discussed in detail in the lectures by Prof. *Riazuddin* that one can show [5–7] that to order α the radiative corrections to β decay are infinite, even if the strong interactions are taken rigorously into account (unless one uses rather specific models [13] for the hadron currents). This prohibits so far a definite check of the universality of the weak interaction. On the practical side there are, however, additional questions of interest: How do the radiative corrections alter the electron spectrum in β decay, or the longitudinal electron polarisation? It is rather remarkable that these properties are largely independent of the details of the strong interactions and can be calculated [14], provided one is willing to neglect contributions of the relative order $\alpha \frac{q}{M}$, $\alpha \frac{E}{M} \ln \frac{M}{E}$, $\left(\text{and for nuclei } \alpha \frac{v_N}{c}\right)$. *Sirlin* [14] has shown that (under these neglections and some general plausible assumptions) all energy dependent corrections to order α can be computed.

The corrections to order α can be depicted as follows:

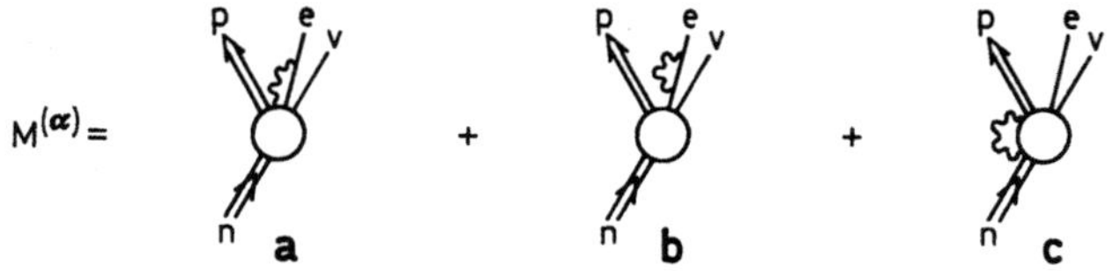

Sirlin splits this into an explicit calculable part which contains all the infrared divergent contributions, is gauge invariant and uv convergent M^c, and another part M^s which contains the details of the strong interactions and which under the above stated neglections contributes only a change of the coupling constants G_V and G_A.

Obviously, graph (*b*) and (under the stated neglections) graph (*c*) lead only to changes in the coupling constants, thus we have only to consider graph (*a*). Sirlin splits $M^{(\alpha)}_{(a)}$ into one part which contains all infrared

* As is the existence of the renormalized perturbation expansion of Q.E.D.

contributions of graph (*a*), and another part which is shown to lead (under the stated neglections) only to changes in the coupling constants, to wit:

$$M^{(\alpha)}_{(a)} = \frac{G}{\sqrt{2}} \frac{\alpha}{4\pi^3} \int \frac{i\,\mathrm{d}^4k\, D^{\mu\nu}(k)}{k^2+2qk+i\varepsilon} \bar{u}_e (2q_\mu + \gamma_\mu \gamma k)\, \gamma^\varrho (1-\gamma_5)\, v_\nu \times \left\{ \frac{2p_\nu - k_\nu}{k^2 - 2pk + i\varepsilon} \langle p | J_\varrho | n \rangle + T_{\nu\varrho}(p, n, k) \right\}. \tag{23}$$

To obtain a gauge invariant result one has to add to the first part of this expression graph (*b*) and part of graph (*c*). Graphically we can depict the whole procedure as follows*:

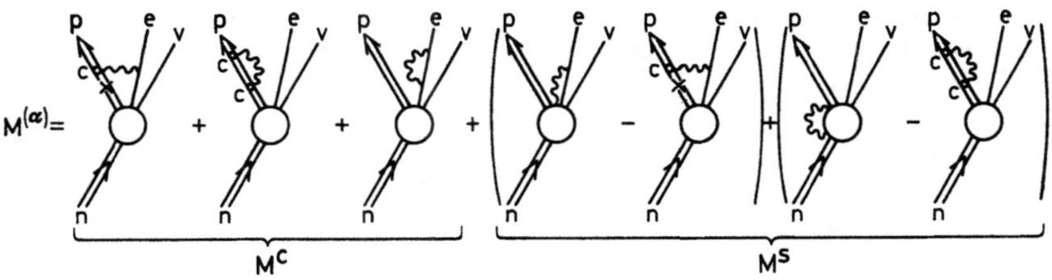

M^c then leads to an energy dependent correction to the electron spectrum of the form:

$$\Delta P^c\, \mathrm{d}^3 q = \frac{\alpha}{2\pi} P^0\, \mathrm{d}^3 q \cdot g(E, E_{\mathrm{max}}, m_e) \tag{24}$$

where P^0 is the spectrum to zero order in α. The function $g(E, E_{\mathrm{max}}, m_e)$ will not be reproduced here, it can be found in *Sirlin*'s paper [14].

The tensor $T_{\nu\varrho}(p, n, k)$ has the property $k^\nu T_{\nu\varrho} = 0$. From this and some regularity assumptions about $T_{\nu\varrho}$ it is possible to show that, under the stated neglections, it contributes only to changes in the coupling constants and thus does not affect the shape of the allowed spectrum.

This finally gives for the radiative corrections to the electron spectrum in low energy β-decay the expression

$$\Delta P\, \mathrm{d}^3 q = \frac{\alpha}{2\pi} P^0\, \mathrm{d}^3 q\, [g(E, E_{\mathrm{max}}, m_e) + C]\,. \tag{25}$$

Only the constant C is so far uncalculable, it is also the only model dependent quantity. The radiative corrections to low energy β-decay are therefore not a very sensitive tool to test specific models.

* The letter *c* in the graphs denotes convection current coupling, the cross on the line means that only the on-shell weak vertex enters.

References

1. *Gell-Mann, M.*: Phys. Rev. **125**, 1067 (1962).
2. *Cabibbo, N.*: Phys. Rev. Letters **10**, 531 (1963).
3. *Eisele, F. et al.*: Z. Phys. **221**, 1 (1969).
4. *Berman, S. M., Sirlin, A.*: Ann. Phys. (N.Y.) **20**, 20 (1962).
5. *Bjorken, J. D.*: Phys. Rev. **148**, 1467 (1967).
6. *Abers, E. S., Norton, R. E., Dicus, D. A.*: Phys. Rev. Letters **18**, 676 (1967). — *Abers, E. S., Dicus, D. A., Norton, R. E., Quinn, H.*: Phys. Rev. **167**, 1461 (1968).
7. *Preparata, G., Weisberger, W. I.*: Phys. Rev. **175**, 1965 (1968).
8. *Adler, S. L.*: Phys. Rev. **177**, 2426 (1969).
9. *Kinoshita, T., Sirlin, A.*: Phys. Rev. **113**, 1652 (1959).
10. *Bardon, M., et al.*: Phys. Rev. Letters **14**, 449 (1965).
11. *Ehrlich, R. D., et al.*: Phys. Rev. Letters **17**, 1118 (1966).
12. See e.g. *Bjorken, J. D., Drell, S. D.*: Quantum Field Theory, Vol. II, 333.
13. *Johnson, K., Low, F. E., Suura, H.*: Phys. Rev. Letters **18**, 1224 (1967). — *Cabibbo, N., Maiani, L., Preparata, G.*: Physics Letters B**25**, 31, B**25**, 132 (1967).
14. *Sirlin, A.*: Phys. Rev. **164**, 1767 (1967).

Univ.-Prof. Dr. *J. Rothleitner*
Institut für Theoretische Physik
der Universität Innsbruck
A-6020 Innsbruck

Unconventional Models of Weak Interactions

GINO SEGRÈ*

Contents

I. Introduction 171
II. Renormalizeable Models of Weak Interactions 172
Model A: i) Leptonic Interactions 172
ii) Semi-leptonic Interactions 174
iii) Non-leptonic Weak Interactions 176
Model B 177
III. Non-Renormalizeable Models 178
Model A: Strongly Interacting *W* Mesons 179
Model B 1: Several *W* Mesons. i) Schizon Models 181
Model B 2: i) Model with Cutoff $\Lambda \sim \frac{1}{\sqrt{G_F}}$ 184
Model C: *W* Mesons and Scalar Mesons with Derivative Coupling 189
IV. Conclusion 190
References 191

I. Introduction

We have already heard a great deal about the so-called conventional theories of weak interactions which are effectively described by evaluating, in lowest order, matrix elements of the interaction Lagrangian

$$\mathscr{L}_{\text{int.}} = \frac{G_F}{2\sqrt{2}} \mathscr{J}_\lambda \mathscr{J}_\lambda^+ + \mathrm{H} \cdot \mathrm{c}. \tag{1}$$

where $\mathscr{J}_\lambda$ is the sum of a hadron current $J_\lambda = V_\lambda + A_\lambda$ and a lepton current [1] j_λ

$$\mathscr{J}_\lambda = J_\lambda + j_\lambda . \tag{2}$$

J_λ is conventionally taken to be the *Cabibbo* [2] current, obtained by a rotation in $SU(3)$ space about the seven axis of a current transforming under $SU(3)$ like a π^+. J_λ and j_λ are given by

$$\begin{aligned} J_\lambda &= \mathrm{e}^{-2i\theta F_7} J_\lambda^{\pi^+} \mathrm{e}^{2i\theta F_7} = J_\lambda^{\pi^+} \cos\theta + J_\lambda^{K^+} \sin\theta , \\ j_\lambda &= i(\bar{e}\gamma_\lambda(1+\gamma_5)\, \nu_e + \bar{\mu}\gamma_\lambda(1+\gamma_5)\, \nu_\mu) \end{aligned} \tag{3}$$

and G_F by $G_F = 10^{-5}/M_N^2$.

* Supported in part by Atomic Energy Commission.

Rather than regard (1) as fundamental, one may choose to consider the weak interactions as being mediated by a charged intermediate vector boson (IVB) of large mass; the fundamental interaction Lagrangian is then

$$\mathcal{L}_{\text{int.}} = g\mathcal{J}_\lambda W_\lambda^+ + \text{H.c.} \tag{4}$$

and the IVB's propagator in momentum space is

$$\Delta_{\mu\nu}(q) = \frac{\delta_{\mu\nu} + \dfrac{q_\mu q_\nu}{M_w^2}}{q^2 + M_w^2}. \tag{5}$$

At momentum transfers $q^2 \ll M_w^2$, we see that calculating to second order in g with (4) reproduces the same results as calculating to first order in G_F with (1) provided

$$\frac{G_F}{\sqrt{2}} = \frac{g^2}{M_w^2}. \tag{6}$$

Differences do show up in the calculation of non-leptonic weak interactions, but we shall say more about that later. It is clear, however, that one runs into a great deal of trouble in trying to calculate anything beyond the lowest order term. We shall not delve on this, as it will be dealt with in other lectures; the difficulty lies in the $q_\mu q_\nu$ term in the W propagator which cause e.g. ladder graphs to diverge like $G_F\left(g^2 \dfrac{\Lambda^2}{M_w^2}\right)^m$ $= G_F(G_F\Lambda^2)^m$ and self mass graphs to behave like $(G_F\Lambda^2)^m$ where Λ is a cutoff. The same asymptotic behaviour holds if we calculate with the $\mathcal{L}_{\text{int.}}$ given in (1). We then may ask the question whether it is possible to have a renormalizeable theory which leads to the same results as 1) or 4) to order G_F, but, by virtue of being renormalizeable, contains no divergences. We shall treat this topic in the next section.

II. Renormalizeable Models of Weak Interactions

Let us preface our remarks by saying there is no renormalizeable model which is completely satisfactory; the best one is probably a model first proposed by *Kummer* and *Segrè* [3] and later rediscovered, and extended by *Christ* [4], whose treatment we shall largely follow.

Model A: i) Leptonic Interactions

We must introduce, in addition to the leptons, two massive (say with mass greater than the K meson) neutral leptons which we call μ_0^* and e_0^*, a charged massive scalar boson $B^\pm$ and a neutral one B^0. The basic

$\mathscr{L}_{\text{int.}}$ will be

$$\begin{aligned}\mathscr{L}_{\text{int.}} = \sqrt{4\pi}\, f\, \{&\bar{e}(1-\gamma_5)\, e_0^* B^- + \bar{\mu}(1-\gamma_5)\, \mu_0^* B^- \\ &+ \bar{\nu}_e(1-\gamma_5)\, e_0^* B^0 + \bar{\nu}_\mu(1-\gamma_5)\, \mu_0^* B^0\} + \text{H.c.}\end{aligned} \tag{7}$$

It is important to note that the hermitean conjugate term which must be added involves $(1+\gamma_5)$ rather than $(1-\gamma_5)$ e.g.

$$(\bar{e}(1-\gamma_5)\, e_0^* B^-)^+ = \bar{e}_0^*(1+\gamma_5)\, e B^+ . \tag{8}$$

The theory is obviously renormalizeable since the mediators of the interaction are scalar bosons. All interactions involving the known leptons such as μ-decay occur to order f^4 and may be readily be cal-

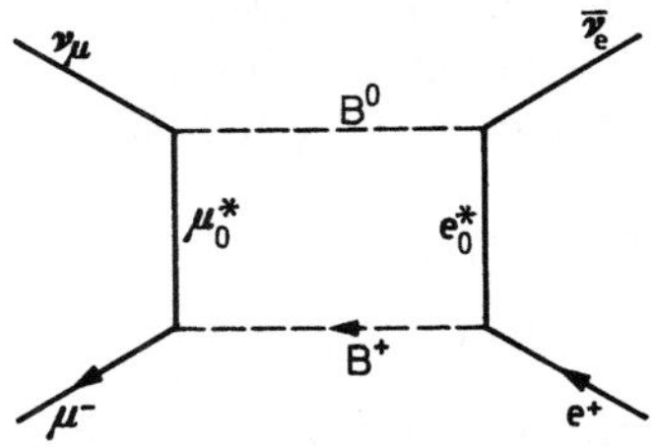

Fig. 1. Lowest order diagram for $\mu^- \to e^- + \nu_\mu + \bar{\nu}_e$

culated; as an example consider μ-decay, which is given in lowest order by the box diagram of Fig. 1.

The matrix element is given, denoting the lepton spinors by u, and the boson masses, which we take as equal, by M

$$\begin{aligned}T = \frac{(4\pi)^2 f^4}{(2\pi)^4} &\int \frac{d^4 q}{q^2 + M^2} \times \frac{1}{(p_{\nu_\mu} - p_\mu - q)^2 + M^2} \\ &\times \bar{u}_e(1-\gamma_5) \frac{(p_e - q + i m_{e_0^*})}{(p_e - q)^2 + m_{e_0^*}^2} (1+\gamma_5)\, u_{\nu_e} \\ &\times u_{\nu_\mu}(1-\gamma_5) \frac{(p_\mu + q + i m_{\mu_0^*})}{(p_\mu + q)^2 + m_{\mu_0^*}^2} (1+\gamma_5)\, u_{\nu_\mu} \\ &= \frac{1}{3} \frac{f^4}{M^2} \bar{u}_e \gamma_\lambda (1+\gamma_5)\, u_{\nu_e} \bar{u}_{\nu_\mu} \gamma_\lambda (1+\gamma_5)\, u_\mu \\ &+ O\left(\frac{m_{e_0^*}^2}{M^2}, \frac{p_l p_{l'}}{M^2}\right)\end{aligned} \tag{9}$$

so that, for M sufficiently large, we obtain the same results for μ decay as (1) or (4) if we set

$$\frac{f^4}{3M^2} = \frac{G_F}{\sqrt{2}} = \frac{g^2}{M_w^2} = \frac{10^{-5}}{M_N^2}. \tag{10}$$

ii) Semi-leptonic Interactions

Let us begin by considering a model in which we take the hadron current to be bilinear in elementary triplet fields p, n and λ. The Cabibbo current, given in (3), would then be written as

$$J_\lambda = i[\cos\theta\, \bar{n}\, \gamma_\lambda(1+\gamma_5)\, p + \sin\theta\, \bar{\lambda}\, \gamma_\lambda(1+\gamma_5)\, p] \tag{11}$$

and we see that results analogous to those of section (*i*) may be obtained for semi-leptonic decays by introducing a neutral heavy fermion f_0 and letting the weak interactions of f_0 and the hadron triplet be given by

$$\begin{aligned}\mathscr{L}_{\text{int.}} = \sqrt{4\pi}\, f\, \{&\bar{p}(1-\gamma_5)\, f_0 B^- \\ &+ (\cos\theta\, \bar{n}(1-\gamma_5)\, f_0 + \sin\theta\, \bar{\lambda}(1-\gamma_5)\, f_0)\, B^0\}\,.\end{aligned} \tag{12}$$

The question is what happens when one turns on the strong interactions, e.g. can one obtain CVC in such a model. One must now evaluate the

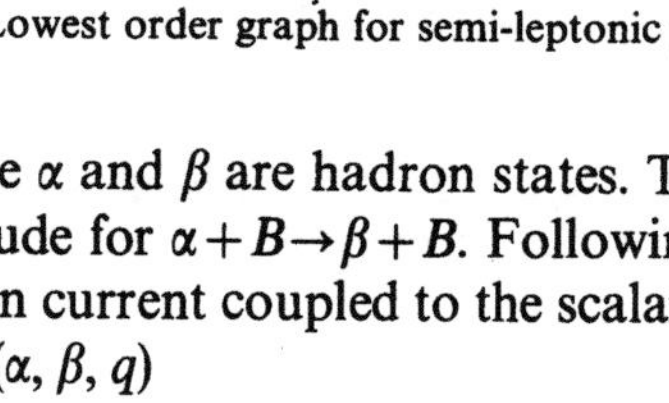

Fig. 2. Lowest order graph for semi-leptonic processes

graph of Fig. 2, where α and β are hadron states. To do this we consider the scattering amplitude for $\alpha + B \to \beta + B$. Following *Christ* [4] we label as S^i the scalar hadron current coupled to the scalar boson B^i and denote the amplitude by $F^{ij}(\alpha, \beta, q)$

$$F^{ij}(\alpha, \beta, q) = i \int e^{iq\cdot y}\, d^4 y \langle\beta|\, T\{S^i(0)\, S^j(y)\}\, |\alpha\rangle\,. \tag{13}$$

Now if we take the f_0 to have no strong interactions, we find, using *Wick*'s theorem

$$\begin{aligned}F^{+0}(\alpha, \beta, q) = &\int e^{iq\cdot y}\, d^4 y \\ &\times \langle\beta|\, \bar{p}(0)\, (1-\gamma_5)\, \Delta(y)\, (1+\gamma_5)\, (n(y)\cos\theta + \lambda(y)\sin\theta)|\, \alpha\rangle\end{aligned} \tag{14}$$

where $\Delta(y)$ is the f_0 propagator. Taking its Fourier transform and making a translation of variables, we find

$$F^{+0}(\alpha, \beta, q) = \frac{4}{(2\pi)^3} \int d^4k\, e^{-ik\cdot y}\, d^4y \frac{1}{(q+k)^2 + M_{f_0}^2} \times \langle \beta | \bar{p}(0)\, (\boldsymbol{q} + \boldsymbol{k})\, (1+\gamma_5)\, (n(y) \cos\theta + \lambda(y) \sin\theta) | \alpha \rangle \,. \tag{15}$$

If we now assume that the integral over k converges very rapidly i.e. that the strong interactions are damped greatly for $k \gg M_n$, we may neglect the k dependence in the propagator for $q \gg M_n$ and obtain

$$\begin{aligned} \lim_{q \gg M_N} F^{+0}(\alpha, \beta, q) &= -i \frac{4(2\pi)^4}{(2\pi)^3} \times \frac{q_\lambda}{q^2 + M_{f_0}^2} \\ &\quad \times \langle \beta | \bar{p}(0) \gamma_\lambda (1+\gamma_5) (\cos\theta\, n(0) + \sin\theta\, \lambda(0)) | \alpha \rangle \\ &= -\frac{8\pi\, q_\lambda}{q^2 + M_{f_0}^2} \langle \beta | J_\lambda^+ | \alpha \rangle \,. \end{aligned} \tag{16}$$

Similarly, for $q \gg M_n$, F^{-0} is proportional to q_λ times the matrix element of J_λ, F^{++}, F^{--} and F^{00} vanish, the latter because of the equality of B^0 and B^{0+}. F^{+-} does not vanish however.

$$F^{+-}(\alpha, \beta, q) = -i \frac{8\pi\, q_\lambda}{q^2 + M_{f_0}^2} \langle \beta | \bar{p}\, \gamma_\lambda (1+\gamma_5)\, p | \alpha \rangle \tag{17}$$

which leads to a contribution of order G_F to $e^- + p \to e^- + p$ or $\mu^- + p \to \mu^- + p$, but not of course to $\nu + p \to \nu + p$ since $B^\pm$ doesn't couple to neutrinos. The former is masked by electromagnetic interactions.

To see in a somewhat different manner how this all comes about, consider that, in the limit large q_0 $F^{ij}(\alpha, \beta, q)$ is given by

$$\lim_{q_0 \to \infty} F^{ij}(\alpha, \beta, q) = \int dy \frac{\langle \beta | [S^i(0), S^j(y, 0)] | \alpha \rangle}{q_0} + O\left(\frac{1}{q_0^2}\right). \tag{18}$$

In the quark model, however, the equal time commutator of two scalar current densities is the fourth component of a vector

$$q^+ [\gamma_4 \lambda^\alpha, \gamma_4 \lambda^\beta] = 2i f^{\alpha\beta\gamma} q^+ \lambda^\gamma q = 2i f^{\alpha\beta\gamma} \bar{q} \gamma_4 \lambda^\gamma q \tag{19}$$

and essentially the same phenomenon holds here. A covariant form of (18) is then

$$\frac{\langle \beta | J_0 | \alpha \rangle}{q_0} \to q_\lambda \frac{\langle \beta | J_\lambda | \alpha \rangle}{q^2} \,. \tag{20}$$

We now calculate the graph of Fig. 2, describing for instance $\alpha \to \beta + e^- + \bar{\nu}_e$

$$T = (4\pi)^2 f^4 \int \frac{d^4 q}{q^2 + M^2} \frac{F^{+0}(\alpha, \beta, q)}{(p_e - p_{\nu_e} - q)^2 + M^2} \times \frac{\bar{u}_e \not{q} (1 + \gamma_5) u_{\nu_e}}{(p_e - q)^2 + m_{e_0^*}^2}. \quad (21)$$

The first term in (18) leads to a matrix element for the semileptonic decay in agreement with (1) and (4) provided we pick f and M as in (10), i.e. such as to have the leptonic decays agree. The other terms in (18) lead to corrections which are

$$O\left(\frac{M_N^2}{M^2}, \frac{M_{f_0}^2}{M^2}\right) \quad (22)$$

and consequently are small, provided M is sufficiently larger than the masses and momenta involved in the problem. The point is then that the equal time commutator for e.g. F^{++} vanishes as do those for F^{--} and F^{00}, while those for F^{+0}, F^{-0} and F^{+-} lead to the results we have already derived.

iii) Non-leptonic Weak Interactions

The great difficulty here is the presence of diagrams such as Fig. 3 which are only semi-weak. An estimate of the characteristic strength of non leptonic $\Delta S = 1$ transitions is

$$\frac{f^2}{4\pi} \sin\theta = \sqrt{\frac{3M^2 G_F}{\sqrt{2}}} \frac{\sin\theta}{4\pi} \sim 3 \times 10^{-4} \frac{M}{M_N} \quad (23)$$

which is almost certainly too large. Christ has proposed ways out of this difficulty by introducing extra particles. The simplest of his models involves two new massive fermions f_0 and f_0' rather than the one we had previously. The interaction Lagrangian is designed so that there are no $\Delta S = 1$ transition to order f^2 and is given by

$$\begin{aligned} \mathscr{L}_{\text{int.}} = \mathscr{L}_{\text{int.}}^{\text{lepton}} + \sqrt{4\pi}\, f \{ &[\bar{p}(1 - \gamma_5) f_0 \cos\theta \\ &+ \bar{p}(1 - \gamma_5) f_0' \sin\theta] B^- + [\bar{n}(1 - \gamma_5) f_0 + \bar{\lambda}(1 - \gamma_5) f_0'] B^0 \} . \end{aligned} \quad (24)$$

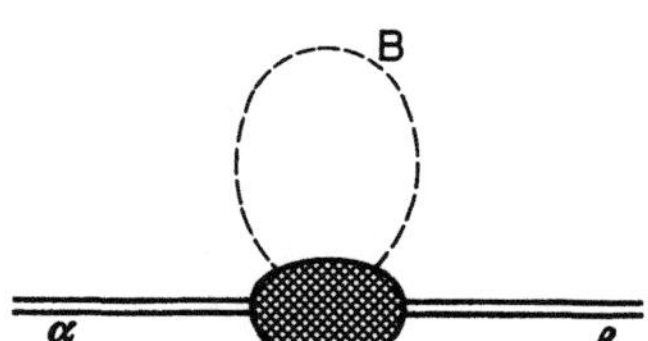

Fig. 3. Lowest order graph for non leptonic processes $\alpha \to \beta$

With the above $\mathscr{L}_{\text{int.}}$, $\Delta S=1$ transition only occur to order f^4 and $\Delta S=2$ transitions to order f^8 so that their magnitude is compatible with the $K_L^0-K_s^0$ mass difference.

There is still a difficulty however and that is the parity violating $\Delta S=0$ transitions, which are still semi-weak and characterized by coupling $f^2/4\pi$. As *Christ* [4] has pointed out however, the nature of the parity violation depends also on the strong interaction Hamiltonian and one may have situations where the parity violation is not incompatible with experiment, despite being semi-weak. We recommend his paper for this and further considerations of this model. Before turning, however, to discussions of other models, we would like to point out that a variation of this model has recently been analyzed in a series of papers by *Shabalin* and co-workers [5]. It differs somewhat in form and has certain new predictions such as the possibility of a sizeable rate for the decay $K\to\pi+\nu+\bar{\nu}$.

Model B

This is a model originally proposed by *Y. Tanikawa* and *S. Watanabe* [6]. Let us describe a simple situation in which we only have four elementary fermions, whose fields we label $\Psi_{1,2,3,4}$. By a Fierz transformation

$$\begin{aligned}&\bar{\Psi}_1(1-\gamma_5)\,\Psi_2\,\bar{\Psi}_4(1+\gamma_5)\,\Psi_3\\&\quad=\tfrac{1}{2}\bar{\Psi}_1\gamma_\lambda(1+\gamma_5)\,\Psi_3\,\bar{\Psi}_4\gamma_\lambda(1-\gamma_5)\,\Psi_2\\&\quad=-\tfrac{1}{2}\bar{\Psi}_1\gamma_\lambda(1+\gamma_5)\,\Psi_3\,\bar{\Psi}_2^{c}\gamma_\lambda(1+\gamma_5)\,\Psi_4^{c}\end{aligned}\tag{25}$$

where the superscript c denotes charge conjugation. Now, letting $\Psi_1\to p$, $\Psi_2\to e^c$ $\Psi_3\to n$ and $\Psi_4\to\bar{\nu}_e^c$ we have

$$\bar{p}(1-\gamma_5)\,e^c\bar{\nu}_e^c(1+\gamma_5)\,n=-\tfrac{1}{2}\bar{p}\gamma_\lambda(1+\gamma_5)\,n\,\bar{e}\gamma_\lambda(1+\gamma_5)\,\nu_e\,. \tag{26}$$

Similarly

$$\bar{\nu}_\mu(1-\gamma_5)\,e^c\bar{\nu}_e^c(1+\gamma_5)\,\mu=-\tfrac{1}{2}\bar{\nu}_\mu\gamma_\lambda(1+\gamma_5)\,\mu\,\bar{e}\gamma_\lambda(1+\gamma_5)\,\nu_e \tag{27}$$

so that if we start with a theory involving scalar bosons coupled to S and PS currents (and hence renormalizeable) with an interaction

$$\begin{aligned}\mathscr{L}_{\text{int.}}\sim ig\{&[\bar{p}(1-\gamma_5)\,e^c+\bar{\mu}(1-\gamma_5)\,\nu_e^c]\,B^0\\&+[\bar{\nu}_\mu(1-\gamma_5)\,e^c+\bar{\mu}(1-\gamma_5)\,\nu_e^c]\,B^+\}+\text{H.c.}\end{aligned}\tag{28}$$

we may obtain the conventional result for neutron β-decay and μ-decay. The theory is, however, very unattractive in that the scalar bosons carry both baryonic and leptonic quantum numbers; it seems extremely difficult furthermore to build into such a model the attractive features of a current theory such as CVC.

In addition to these two models, there is one by *Fronsdal* [7], which he claims may be renormalizeable and most recently, a very interesting analysis by *T. D. Lee* [8] of the possibility of removing all the divergences in weak interactions by use of an indefinite metric. If time allows we will return to this question at the end of our lectures.

III. Non-Renormalizeable Models

There are several unconventional non-renormalizeable models of weak interactions. The main problem they address themselves to is the question of divergences caused by higher order weak interactions; since this question will be treated in other lectures, we will not go into any detail of the nature of these divergences in a conventional model. We do see that, for instance, the exchange of two W mesons, as in Fig. 4, between α and β (which can be multiparticle states) leads to a quadratic divergence.

If we introduce a cutoff Λ, the characteristic size of the matrix element $\alpha \to \beta$ is $G_F(G_F \Lambda^2)$. If we now try to set an upper limit on Λ by comparing the computed matrix element with the known $K_L^0 - K_S^0$ [3, 10] mass difference, we find $\Lambda \sim 3$–4 BeV. From semi-leptonic decays, the best limit on $\Lambda \sim 50$ BeV is obtained by computing [9–11] the matrix element for the as of yet unobserved process $K_L^0 \to \mu^+ \mu^-$.

The question that comes immediately to mind is what provides a cutoff as low as 3–4 BeV. In particular since, if the lower limit on the W mass is ~ 2 BeV, does it make any sense to have a cutoff of integrals over W momenta which is presumably of the same order of magnitude as the W mass? One can of course go a step further and ask does it make any sense in general to compare with experiment a given order of a series, all of whose terms diverge, neglecting higher order terms. For instance

$$\lim_{x \to \infty} \frac{1}{1-x} = 0 \tag{29}$$

and it obviously doesn't make sense to expand

$$\frac{1}{1-x} = 1 + x + \cdots \tag{30}$$

and then set a limit on how large x can be by stopping the expansion at the first term. With that as a warning, let us go on to describe a few models.

Model A: Strongly Interacting *W* Mesons

In such a model, it is hoped that W meson strong interactions will provide the rapid damping necessary to provide a low cutoff. One can distinguish between models in which the W mesons have only strong interactions among themselves [12, 13] such as λW^4 interaction and others in which they interact strongly with hadrons [14–16]. Very little can be said about the former type of model, while the latter would appear to be inconsistent with experiment. In this section we would like to show how this need not necessarily be true.

We shall follow *Callan*'s paper [15], in letting the total Lagrangian which includes the interaction term (4) be modified by adding a new term

$$\mathscr{L} \to \mathscr{L}' = \mathscr{L} - \frac{F_{\mu\nu}^{+} F_{\mu\nu}}{4} R \tag{31}$$

where $F_{\mu\nu} = \partial_\mu W_\nu - \partial_\nu W_\mu$ and R is a scalar object constructed out of hadron fields, involving no derivatives of fields and invariant under $SU(3) \times SU(3)$. $\mathscr{L}'$ is given by

$$\begin{aligned}\mathscr{L}' = & -\tfrac{1}{4} F_{\mu\nu}^{+} F_{\mu\nu}(1+R) - \tfrac{1}{2} M_w^2 W_\mu^{+} W_\mu \\ & + g(J_\mu W_\mu^{+} + J_\mu^{+} W_\mu) + \mathscr{L}_{\text{hadron}} + \mathscr{L}_{\text{lepton}}\end{aligned} \tag{32}$$

where $\mathscr{L}_{\text{hadron(lepton)}}$ represents the part of $\mathscr{L}$ involving only hadrons (leptons).

If we label the hadron fields as $\varphi_j(x)$ and consider an infinitesimal local gauge transformation [1]

$$\varphi_j(x) \to \varphi_j(x) + \tau(x)\, f_j(\varphi(x)) \tag{33}$$

we see that, to first order in τ

$$\delta\mathscr{L} = \frac{\delta\mathscr{L}}{\delta\tau}\tau + \frac{\delta\mathscr{L}}{\delta(\partial_\mu \tau)}\partial_\mu \tau \tag{34}$$

with

$$\frac{\delta\mathscr{L}}{\delta\tau} = \sum_j \left\{ \frac{\delta\mathscr{L}}{\delta\varphi_j} f_j + \frac{\delta\mathscr{L}}{\delta(\partial_\mu \varphi_j)} \partial_\mu f_j \right\}. \tag{35}$$

The current J_μ is

$$J_\mu = -\sum_j \frac{\delta\mathscr{L}}{\delta(\partial_\mu \varphi_j)} f_j = -\frac{\delta\mathscr{L}}{\delta(\partial_\mu \tau)} \tag{36}$$

and its divergence is obtained, by using (35) and the Euler-Lagrange equation

$$\frac{\delta\mathscr{L}}{\delta\varphi_j} = \partial_\mu \frac{\delta\mathscr{L}}{\delta\partial_\mu \varphi_j} \tag{37}$$

and equals then

$$\partial_\mu J_\mu = -\frac{\delta \mathscr{L}}{\delta \tau}. \tag{38}$$

Since the extra term in (31) is invariant under $SU(3) \times SU(3)$ and contains no derivatives of fields we see that the currents J_μ have the same commutation relations and divergences as they had before.

The matrix elements for semi-leptonic processes $\alpha \to \beta + l + \bar{\nu}_e$ are proportional to

$$g\bar{u}_e \gamma_\mu (1+\gamma_5)\, u_{\bar{\nu}_e} \langle \beta | W_\mu | \alpha \rangle \tag{39}$$

where the matrix element of W_μ can be determined from the equations of motion which read

$$\partial_\mu (F_{\mu\nu}(1+R)) + M_w^2 W_\nu = g J_\nu \tag{40}$$

so that

$$\langle \beta | W_\mu | \alpha \rangle = \frac{g}{M_w^2} \langle \beta | J_\mu | \alpha \rangle + \frac{\langle \beta | \partial_\nu (F_{\nu\mu}(1+R)) | \alpha \rangle}{M_w^2}. \tag{41}$$

At $q=0$ only the first term survives and we have the usual predictions of the Cabibbo theory; for $q \neq 0$ the second term contributes and we would expect appreciable deviations from (4) for sufficiently large momentum transfers.

Lepton-hadron universality holds in this model because the W propagator $P_{\mu\nu}(q)$ satisfies

$$P_{\mu\nu}(0) = \frac{\delta_{\mu\nu}}{M_w^2}. \tag{42}$$

To see this consider

$$P_{\mu\nu}(q) = \int d\mu^2 \sigma(\mu^2) \frac{\left(\delta_{\mu\nu} + \dfrac{q_\mu q_\nu}{\mu^2}\right)}{q^2+\mu^2} \tag{43}$$

where $\sigma(\mu^2)$ is defined by the two point function

$$\begin{aligned} &\langle 0 | W_\mu^+(x)\, W_\nu(0) | 0 \rangle \\ &= \frac{1}{(2\pi)^3} \int d^4 p\, \theta(p_0)\, e^{-ipx} \left(\delta_{\mu\nu} + \frac{p_\mu p_\nu}{p^2}\right) \sigma(-p^2) \\ &= \frac{1}{(2\pi)^3} \int d^4 p\, d\mu^2\, \delta(p^2+\mu^2)\, \sigma(\mu^2)\, \theta(p_0)\, e^{-ipx} \left(\delta_{\mu\nu} + \frac{p_\mu p_\nu}{p^2}\right). \end{aligned} \tag{44}$$

From the equation of motion (40), we see that

$$W_0 = \frac{g J_0 - \partial_k (F_{0k}(1+R))}{M_w^2} \tag{45}$$

but

$$\frac{\partial \mathcal{L}}{\partial W_K} = F^{+}_{0K}(1+R) \tag{46}$$

is the canonical momentum conjugate to W_K where $K=1,2,3$ so that

$$\delta(x_0)\langle 0|[W^{+}_{0}(\boldsymbol{x},0), W_K(0)]|0\rangle = -\frac{\partial_K \delta^4(x)}{M^2_w}. \tag{47}$$

Comparing this with the spectral form we obtain

$$\int \frac{\sigma(\mu^2)}{\mu^2} \mathrm{d}\mu^2 = \frac{1}{M^2_w} \tag{48}$$

from which (42) and hence lepton hadron universality follow.

Decays which are forbidden in lowest order such as $K^0_L \to \mu^+\mu^-$ are still forbidden because R conserves strangeness so that we do not have a strong matrix element connecting K^0_L to a W^+W^- state. The form of the interaction also ensures that one doesn't produce single W mesons strongly, only in pairs. The fact that one hasn't seen such a pair produced strongly does suggest however that the W mass must be rather large. It would probably be interesting to estimate what the limits are for such a case, though it's hard to calculate the strong interaction effects.

The theory of non leptonic interaction is much the same as in the usual picture. A priori one might expect difficulties since e.g. for large q

$$P_{\mu\nu}(q) \sim \int \mathrm{d}\mu^2 \sigma(\mu^2)\left(\delta_{\mu\nu} + \frac{q_\mu q_\nu}{\mu^2}\right)\Big/ q^2 \sim \frac{1}{q^2}\left\{\int \mathrm{d}\mu^2 \sigma(\mu^2) + \frac{q_\mu q_\nu}{M^2_w}\right\} \tag{49}$$

and it is not clear that the integral over $\sigma(\mu^2)$ converges. However it is in the spirit of the model to assume that the strong interactions act to cut off all integrals at low values of q. For further discussions of such models, including the possibility of CP violation see Ref. [13].

Model B 1: Several *W* Mesons. i) Schizon Models

The main thrust of these models is directed to obtaining the experimentally observed octet enhancement of non leptonic decays. In our basic interaction Lagrangian, the hadronic part has the form

$$\mathcal{L}^{\text{hadr}}_{\text{int.}} = \frac{G_F}{2\sqrt{2}}(J^{\pi^+}_\lambda \cos\theta + J^{K^+}_\lambda \sin\theta)(J^{\pi^-}_\lambda \cos\theta + J^{K^-}_\lambda \sin\theta) + \text{H.c.} \tag{50}$$

Since the currents belong to the eight representation $\mathcal{L}^{\text{hadr.}}_{\text{int.}}$ has both an eight and a twenty-seven component, whereas experimentally only the former is seen. To remedy this one can resort to a dynamical enhancement

[17] or to adding other W mesons so that the order G_F effective Lagrangian will transform like an octet component. There are several ways of doing this [18–21] and we shall only describe the scheme of Ref. [18]. The $SU(3)$ hadron currents may have their $SU(3)$ transformation properties labelled by a pair of indices $i, j = 1, 2, 3$ where the convention is to take T_j^i to represent a singlet and an octet. If we denote by $S_i(S^i)$ a tensor transforming like the 3 $(\bar{3})$ representation of $SU(3)$, then $S^i S_i$ is a reducible nonet comprising an octet

$$T_i^j = S^j S_i - \tfrac{1}{3}\delta_i^j S_k S_k \tag{51}$$

and a singlet $O = S^k S_k$ where we sum over repeated indices. To illustrate the labelling we write down the 3×3 traceless matrix representing the pseudoscalar mesons

$$M = \begin{pmatrix} \dfrac{\eta}{\sqrt{6}} + \dfrac{\pi^0}{\sqrt{2}} & \pi^+ & K^+ \\ \pi^- & \dfrac{\eta}{\sqrt{6}} - \dfrac{\pi^0}{\sqrt{2}} & K^0 \\ K^- & \overline{K^0} & -\sqrt{\dfrac{2}{3}}\,\eta \end{pmatrix}. \tag{52}$$

The D'Espagnat Lagrangian involves an $SU(3)$ triplet of W mesons, which we label as W^i, $i = 1, 2, 3$ and the hadronic part of the interaction Lagrangian may be written as

$$\mathscr{L}_{\text{int.}}^{\text{had.}} = g\{\cos\theta\, J_i^2 W^i + \sin\theta\, J_i^3 W^i\} + \text{H.c.} \tag{53}$$

where $(J_j^i W^j)^+ = W_j J_i^j$ and g is defined as in (6). If we kept only the charged W meson, W^1, we would have the usual theory, for which the effective non-leptonic $\mathscr{L}_{\text{int.}}$ behaves as

$$\sim \frac{G_F}{\sqrt{2}}\{\cos^2\theta\, J_1^2 J_2^1 + \sin^2\theta\, J_1^3 J_3^1 + \cos\theta\sin\theta (J_1^2 J_3^1 + J_1^3 J_2^1)\} + \text{H.c.} \tag{54}$$

whereas for the D'Espagnat Lagrangian it goes as

$$\begin{aligned} &\sim \frac{G_F}{\sqrt{2}}\{\cos^2\theta\, J_i^2 J_2^i + \sin^2\theta\, J_i^3 J_3^i + \cos\theta\sin\theta (J_i^2 J_3^i + J_i^3 J_2^i)\} \\ &= \frac{G_F}{\sqrt{2}}\{\cos^2\theta\, H_2^2 + \sin^2\theta\, H_3^3 + \sin\theta\cos\theta (H_3^2 + H_2^3)\}, \end{aligned} \tag{55}$$

i.e. the $\Delta S = 1$ effective Lagrangian transforms like the sixth component (K_s^0) of an octet and the $\Delta S = 0$ part has singlet and octet components.

In attempting to formulate a model of weak interactions with neutral W's, one is immediately struck by the unappealing fact that these W's appear to couple only to hadrons, as there is no experimental evidence for neutral leptonic currents. An attempt to have such currents in the Lagrangian was made by *Good, Michel*, and *de Rafael* [22]. They denote by l^1 and l^2 the fields creating e^- and ν_e. Then $l_1=(l^1)^+$ and $l_2=(l^2)^+$ create e^+ and $\bar{\nu}_e$. For muons they make the opposite choice namely L_1, L_2, L^1 and L^2 create respectively $\nu_\mu, \mu^-, \bar{\nu}_\mu$ and μ^+. They couple the leptons to the W muons as follows

$$\begin{aligned}\mathscr{L}_{\text{int.}}^{\text{leptons}} = &= g\varepsilon\{l^2 l_1 W^1 + l^2 l_2(W^2\cos\theta + W^3\sin\theta)\\ &+ L^2 L_1 W^1 + L^2 L_2(W^2\cos\theta + W^3\sin\theta)\}\end{aligned} \tag{56}$$

where θ is the Cabibbo angle and $|\varepsilon|=1$ and then show that with a choice of $\varepsilon=i$, there are no neutral lepton current-hadron current couplings to order G_F.

The model is certainly very elegant; one can in fact rewrite it in an even more streamlined form by using the concept of the U spin subgroup of $SU(3)$.

$$\begin{aligned}U_{\pm} &= \tfrac{1}{2}(F_6 \pm iF_7),\\ U_3 &= \tfrac{1}{4}(\sqrt{3}F_8 - F_3).\end{aligned} \tag{57}$$

U spin generators commute with the electromagnetic Hamiltonian and hence representations of the U spin group have the same charge, e.g. π^+ and K^+ form a U spin doublet. Now the weak $\mathscr{L}_{\text{int.}}^{\text{hadr.}}$ given in (53) transforms like a linear combination of the two neutral members of an $SU(3)$ triplet; they in turn form a U spin doublet. One may then go to a rotated frame (obtained by rotating through 2θ about the second axis in U spin space) and there rewrite (53) as

$$\mathscr{L}_{\text{int.}}^{\text{hadron}} = gJ_i'^2\, W'^i \tag{58}$$

where J' and W' are the rotated currents and W fields.

This model does not say anything new about the divergence problem. In fact, if anything, the cutoff Λ must be even smaller than in the conventional theory. Furthermore it has been claimed by McKellar [23] that the $\Delta S=0$ parity violating part of the Hamiltonian is too large to be compatible with the data on circular polarization of γ rays from heavy nuclei. The parity violating part of the one pion exchange force is calculated by first evaluating the weak parity violating $NN\pi$ vertex. In the soft pion limit this is proportional to:

$$\langle N|[Q_5, \mathscr{L}_{\text{int.}}^{\text{p.v.}}]|N\rangle\,. \tag{59}$$

Only the $I=1$ part of $\mathscr{L}_{\text{int.}}$ contributes since

$$[Q_5, \mathscr{L}_{\text{int.}}^{\text{p.v.}}] = a_1 \mathscr{L}_{\text{int.}}^{\text{p.c.}}\big|_{I=1} + a_2 \mathscr{L}_{\text{int.}}^{\text{p.c.}}\big|_{I=2} \tag{60}$$

and only the $I=1$ component has non zero matrix elements between nucleon states. In a conventional theory, we see that the $I=1$ part of the effective Lagrangian is proportional to $\sin^2\theta$ since

$$J_1^2 J_2^1 \sim J^{\pi^+} J^{\pi^-} \tag{61}$$

and this only has, upon symmetrizing, $I=0, 2$. In the D'Espagnet model, on the other hand H_2^2 has an $I=1$ component, so that the $\Delta I=1$ parity violating pion exchange force differs by a factor $\sim \sin^2\theta$ in the two model and *McKellar* [23] claims he can rule out the D'Espagnet model on these grounds.

Model B 2: i) Model with Cutoff $\Lambda \sim \dfrac{1}{\sqrt{G_F}}$

a) *Selection Rules and Decay Rates*

We would now like to discuss a model recently proposed by the present author [24]. It involves three W mesons, of which two, $W_a^\pm$ and $W_b^\pm$ are charged and the third, $W_c^{0,\bar{0}}$ is neutral but not self conjugate. It also assumes the existence of two neutral massive ($> M_k$) leptons which we call λ_e and λ_μ so that we have two lepton triplets (e^-, ν_e, λ_e) and $(\mu^-, \nu_\mu, \lambda_\mu)$. If we define $\tilde{j}_\tau$ analogously to the j_τ of (3)

$$\tilde{j}_\tau = i\bar{e}\,\gamma_\tau(1+\gamma_5)\,\lambda_e + i\bar{\mu}\,\gamma_\tau(1+\gamma_5)\,\lambda_\mu \tag{62}$$

we may write $\mathscr{L}_{\text{int.}}$, dropping for the moment the Lorentz indices, as

$$\begin{aligned}\mathscr{L}_{\text{int.}} = g\,\{&J^{\pi^+} W_a^- + J^{K^+} W_b^- \\ &+ (J^{K^0} + \gamma V^0)\, W_c^0 + j(W_a^- \cos\theta + W_b^- \sin\theta) \\ &+ \tilde{j}(-W_a^- \sin\theta + W_b^- \cos\theta)\} + \text{H.c.}\end{aligned} \tag{63}$$

where J^i is the usual hadronic $V-A$ current, V^0 is the conserved vector $SU(3)$ singlet current, whose fourth component integrated over all space is the baryonic charge and γ is an adjustable parameter. The effective Lagrangian, to order G_F, is given by

$$\begin{aligned}\mathscr{L}_{\text{eff.}} = \frac{G_F}{\sqrt{2}}\,\{&J^{\pi^+} J^{\pi^-} + J^{K^+} J^{K^-} \\ &+ (J^{K^0} + \gamma V^0)(J^{\overline{K^0}} + \gamma V^0) + J^{\pi^+} j^+ \cos\theta \\ &+ J^{K^+} j^+ \cos\theta - J^{\pi^+} \tilde{j}^+ \sin\theta + J^{K^+} \tilde{j}^+ \cos\theta \\ &+ jj^+ + \tilde{j}\tilde{j}^+\}\,.\end{aligned} \tag{64}$$

It is clear that the selection rules and rates of the conventional theory follow from the above since $\Delta S=0$ semileptonic decays are proportional to $\cos\theta\, J^{\pi^+} j^+$, $\Delta S=1$ semi-leptonic decays to $\sin\theta\, J^{K^+} j^+$ and purely leptonic decays to jj^+. The fact that $\lambda_{e,\mu}$ have masses greater than m_K insures they are not produced in semi-leptonic decays. Furthermore there is no $j\tilde{j}^+$ term so that even if e.g. λ_μ were more massive than λ_e, a decay such as $\lambda_\mu \to \lambda_e + \mu^- + e^+$ would not take place.

Non leptonic $\Delta S=1$ decays occur only through the interference term $\gamma(J^{K^0} V^0 + V^0 J^{\overline{K^0}})$ and hence are purely octet since V^0 is an $SU(3)$ singlet. $\gamma=1$ corresponds to an enhancement of $\Delta S=1$ non leptonic over $\Delta S=1$ semi leptonic decay rates by a factor $1/\sin^2\theta$ in the absence of strong interactions. We might think that McKellar's objection to the D'Espagnat model also holds here, but such is not the case since

$$J^{K^+} J^{K^-} + J^{K^0} J^{\overline{K^0}} + \text{H.c.} \tag{65}$$

is an $SU(2)$ scalar, i.e. has no $I=1$ component.

b) *Divergence in Leptonic and Semi-Leptonic Processes*

Let us begin by examining the divergences in $\Delta S=1$ semi-leptonic decays, as in e.g. Fig. 4 with $\alpha = K_L^0$ and $\beta = \mu^+\mu^-$. In this model there are two diagrams contributing, as shown in Fig. 5.

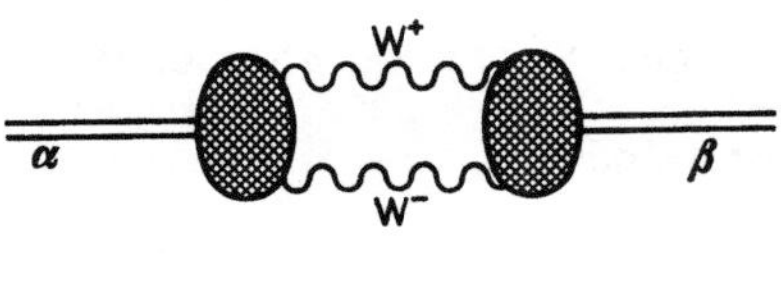

Fig. 4. Two W meson exchange between states α and β

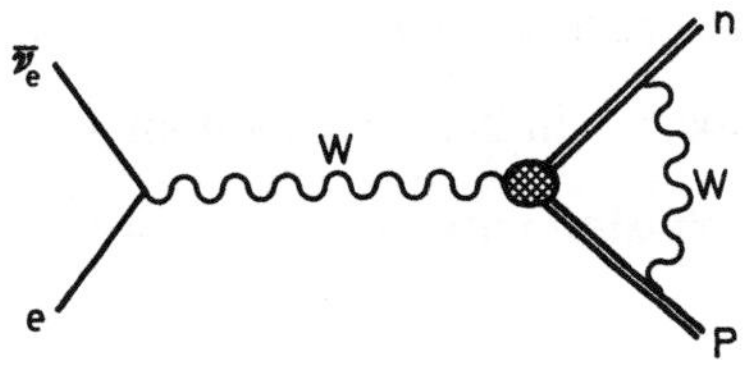

Fig. 5. Lowest order diagrams contributing to $K_L^0 \to \mu^+\mu^-$

Each one of the diagrams diverges quadratically, the first being proportional to $G_F^2 \Lambda^2 \sin\theta\cos\theta$ and the second to $-G_F^2\Lambda^2 \sin\theta\cos\theta$, so that the sum has no quadratic divergence. Instead we find only a logarithmic divergence proportional to $G^2 m_{\lambda_\mu}^2$. The reason for this [4] is that if $m_{\lambda_\mu}=0$, we would be able to rewrite the Lagrangian in terms of new fields ν'_μ, λ'_μ

$$\begin{aligned} \nu_\mu &= \nu'_\mu \cos\theta + \lambda'_\mu \sin\theta\,, \\ \lambda_\mu &= -\nu'_\mu \sin\theta + \lambda'_\mu \cos\theta \end{aligned} \tag{66}$$

as

$$\mathcal{L}_{\text{int.}}^{\text{lepton}} = g\{j' W_a^- + \tilde{j}' W_b^-\}\,. \tag{67}$$

We would assign $S=0$ to μ^-, ν'_μ and W_a, $S=-1$ to λ'_μ and W_b and there would be no strangeness non conservation except for the W_c coupling. Hence (to lowest order) $K_L^0 \nrightarrow \mu^+\mu^-$; in our case the matrix element must be proportional to m_{λ_μ} to reflect this and is only logarithmically divergent. Even if we go to higher order these $\Delta S=1, \Delta Q=$ processes remain small, that is higher order processes contribute so the matrix element becomes

$$T_{K_L\to\mu^+\mu^-} \sim G_F^2 m_{\lambda_\mu}^2 \sum_{m=0}^{\infty} a_m (G_F \Lambda^2)^m \tag{68}$$

where a_m is a function of the momenta and couplings of the external particles. We see that $G_F\Lambda^2 \sim 1$ is consistent with experimental limits.

This cancellation of divergences does not hold for purely leptonic or $\Delta S = \Delta Q = 0$ semi-leptonic processes, but before talking about them we should specify more carefully our expansion parameter. Every internal W meson leads to a quadratic divergence because of the $q_\mu q_\nu$ part of the propagator: calling C the matrix element of the commutators of currents, we have a divergence which goes as

$$\frac{g^2}{2(2\pi)^4} \times \frac{C}{M_w^2} \int \frac{d^4 q}{q^2 + M_w^2} \sim \frac{C G_F \Lambda^2}{\sqrt{2} \times 16\pi^2} \tag{69}$$

where the integral over q^2 has been cut off at $q^2 = \Lambda^2$. C is typically of the order of four so our characteristic expansion parameter $\sim \dfrac{G_F \Lambda^2}{4\pi^2}$. This is similar to quantum electrodynamics where the characteristic expansion parameter is $\dfrac{\alpha}{\pi} = \dfrac{e^2}{4\pi^2}$. Here we have

$$\frac{g^2}{4\pi^2}\left(\frac{\Lambda^2}{M_w^2}\right).$$

Therefore a process such as $\nu_\mu + P \to \nu_\mu + P$, although not forbidden in this model, even for $G_F\Lambda^2 \sim 1$ has a small matrix element with respect to $\nu_\mu + n \to \mu^- + p$

$$\frac{\text{Rate}\,(\nu_\mu + p \to \nu_\mu + p)}{\text{Rate}\,(\nu_\mu + n \to \mu^- + p)} \sim \left(\frac{1}{4\pi^2}\right)^2 \sim 5 \times 10^{-4} \tag{70}$$

whereas the experimental limit is that the ratio is less than 5×10^{-2} [25]. The most severe limit on Λ in leptonic and semileptonic decays in this model is lepton-hadron universality, namely deviations from $G_\mu = G_\beta \cos\theta$ due to vertex corrections as depicted in Fig. 6. Of course μ decay has corresponding vertex corrections, which tend to balance off those of neutron β decay. The main difference is that W_c^0 couples to hadrons, but not to leptons so we have a non zero deviation from $G_\mu = G_\beta \cos\theta$. Calculations show, however, that for $G_F \Lambda^2 \sim 1$, the deviation is $\sim < 1\%$ and hence compatible with experiment.

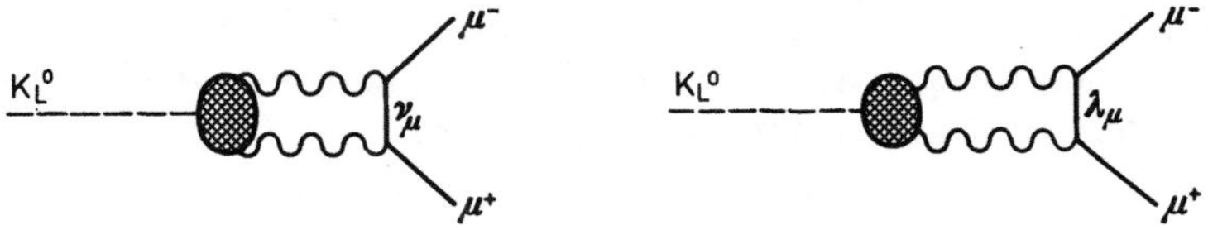

Fig. 6. Vertex correction to neutron β-decay

c) *Divergences in Non Leptonic Processes*

In an IVB model, the G_F term is formally quadratically divergent. The matrix element for $\alpha \to \beta$ is

$$T_{\alpha\to\beta} = \frac{ig^2}{2(2\pi)^4} \int \frac{d^4 q}{q^2 + M_w^2} \left(\delta_{\mu\nu} + \frac{q_\mu q_\nu}{M_w^2}\right) \times \int d^4 x \, e^{-iq\cdot x} \langle \beta | T\{J_\mu^+(x) \, J_\nu(0)\} | \alpha \rangle . \tag{71}$$

The quadratic divergence arises of course from the $q_\mu q_\nu$ term, and is given by

$$T_{\alpha\to\beta}\big|_{\text{quad. div.}} = -i \frac{G_F \Lambda^2}{\sqrt{2} \times 16\pi^2} \int d^4 x \, e^{iq\cdot x} \, \delta(x_0) \times \langle \beta | [J_0^+(x), \partial_\nu J_\nu(0)] | \alpha \rangle . \tag{72}$$

The evaluation of the commutator in conventional models will be discussed in other lectures. It is necessary to know the form of the Hamiltonian in order to be able to calculate the divergence of the current; the commutator in (72) is thus model dependent. All one really needs to know of course is the part of the Hamiltonian that breaks $SU(3) \times SU(3)$ since the currents are conserved in the symmetry limit

$$[J_0, H] = \left[J_0, H_{\substack{SU(3)\times SU(3) \\ \text{breaking}}}\right] = \partial_0 J_0 \to \partial_\mu J_\mu . \tag{72a}$$

We take for H the form

$$H = H_0 + \varepsilon_0 u_0 + \varepsilon_8 u_8 \tag{73}$$

where H_0 is invariant under $SU(3) \times SU(3)$ and u_0 and u_8 are components of a scalar nonet belonging to the $(3, \bar{3}) + (\bar{3}, 3)$ representation of $SU(3) \times SU(3)$.

$$\begin{aligned} \partial_\mu V^i_\mu &= i\varepsilon_8 f^{8ij} u_j\,, \\ \partial_\mu A^i_\mu &= -i(\varepsilon_0 d^{0ij} v_j + \varepsilon_8 d^{8ij} v^j) \end{aligned} \tag{74}$$

where the v_j are a nonet of pseudoscalar densities. Since α and β are hadron states, if the commutator in (72) contained terms which violated e.g. parity, strangeness, isospin, we would have apparent large violations of parity etc. since $G_F \Lambda^2$ is taken to be of order one.

In our model however, the $\Delta S = 1$ part of the commutator is given by

$$[V^0_0(x),\, \delta_\mu J^{K_0}_\mu(0)]\, \delta_\mu(x_0) = 0 \tag{75}$$

since the baryonic number commutes with the $SU(3) \times SU(3)$ currents and with their divergences in this model. Hence there is no quadratically divergent term to order G_F. The $\Delta S = 0$ part is more complicated; one has to evaluate the commutator

$$\sum_{i=1,2,4,5,6,7} [J^i_0(x),\, \partial_\mu J^i_\mu(0)]\, \delta(x_0)\,. \tag{76}$$

Surprisingly enough, the commutator is proportional to $\varepsilon_0 u_0 + \varepsilon_8 v_8$ i.e.

$$T_{\alpha\to\beta}\big|_{\Delta S=0\,\substack{\text{quad.}\\ \text{div.}}} = \frac{G_F \Lambda^2}{\sqrt{2} \times 4\pi^2} \langle \beta | \varepsilon_0 u_0(0) + \varepsilon_8 u_8(0) | \alpha \rangle \tag{77}$$

which just corresponds to a form preserving addition to the semistrong Hamiltonian. It is obviously compatible with $G_F \Lambda^2 \sim 1$. Estimates of higher order in G_F terms are of course very complicated; to obtain a $\Delta S = 1$ transition one will always need to bring in a W_c meson coupled at one end to J^{K^0} and at the other to V^0. Since the latter commutes at equal times with all $SU(3) \times SU(3)$ currents and their divergences, it appears that $\Delta S = 1$ transitions will never be maximally divergent i.e. of the form $(G_F \Lambda^2)^n$. $\Delta S = 0$ processes also continue to have attractive features in higher order, such as no parity violation in maximally divergent terms etc.; for a complete discussion of these processes in this model see Ref. [27].

$\Delta S = 2$ transitions are also more convergent in this model because to obtain a non zero matrix element for such a process, we need to couple two W_c mesons to the hadrons through V^0 at one end and J^{K_0} at the other. Once again we find that the matrix element is more convergent than in the usual theory because of the vanishing of the commutators of V^0 and other currents or divergences.

We hasten to add however that neither in this model, for which $G_F \Lambda^2 \sim 1$, nor in other models which remove divergences, can the IVB mass be too large. The reason is that the coupling constant g^2 is fixed by (6) and if M_w is too large, higher order weak interactions will be important even if they aren't divergent for the simple reason that $g^2/4\pi$ will be so large. It is of course very hard to set an upper limit on M_w this way. It is probably once again the $K_L - K_S$ mass difference which places the most severe restriction; it would be surprising if the IBV mass were larger than 5–10 BeV [28], though we emphasize it is hard to make a reliable estimate. If $M_w \sim 100$ BeV, we would have $g^2/4\pi \sim \alpha$, the fine structure constant, and this is certainly too large.

Model C: *W* Mesons and Scalar Mesons with Derivative Coupling

We shall say only a few words about a model proposed by *Gell-Mann et al.* [29], which has several very attractive features. The basic idea is to have weak currents couple not only to *W* mesons, but to scalar mesons with a derivative coupling in such a way as to cancel the contribution of the $q_\mu q_\nu$ part of the propagator to a large class of matrix elements. The idea here is to write the weak interaction Lagrangian as

$$\mathscr{L}_{\text{int.}} = g \sum_i \mathscr{J}_{\lambda}, : B_{\lambda}, : \tag{78}$$

where i are a set of indices. If we let $B_{\lambda,i} = \alpha_i W_\lambda$, the theory becomes the same as that of (4) provided we choose the constant α_i in the same manner. If we take instead, e.g.

$$B_{\lambda,i} = \alpha_i W_\lambda + \frac{1}{\tau_i} \partial_\lambda \varphi_i \tag{79}$$

where φ_i are scalar meson fields, corresponding to particles which interact weakly with hadrons as do the W's, and τ_i's are constants. The effective weak interaction propagator is now given by the Fourier transform of

$$\langle 0 | T\{B^+_{\lambda,i}(x)\, B_{\sigma,j}(y)\} | 0 \rangle \tag{80}$$

and one may arrange constants such that the propagator $\Delta_{\lambda\sigma,ij}$ (9) tends to zero at least like $1/q^2$ for $i \neq j$. One may thus eliminate divergences in semi-leptonic and strangeness changing non leptonic decays, in parity violating $\Delta S = 0$ transitions etc. For $i = j$, however, the divergences add, and there is no way one can get around this as long as the Hamiltonian remains Hermitian. Consider for the moment a_μ, a neutral vector meson of mass M, and b, a neutral scalar field of mass M_2. If we let

$$\phi_\lambda = a_\lambda + \frac{1}{M_i} \partial_\lambda b \tag{81}$$

the Fourier transform of the vacuum expectation value of the time ordered produce of $\phi_\lambda(X)$ and $\phi_\sigma(0)$ is

$$\Delta_{\lambda\sigma}(q) = \frac{\delta_{\lambda\sigma} + \dfrac{q_\lambda q_\sigma}{M_1^2}}{q^2 + M_1^2} + \frac{\dfrac{q_\lambda q_\sigma}{M_1^2}}{q^2 + M_2^2}. \tag{82}$$

While if we define

$$\begin{gathered} \tilde{\phi}_\lambda = a_\lambda + \frac{i}{M_1} \partial_\lambda b, \\ \tilde{\Delta}_{\lambda\sigma}(q) = \frac{\delta_{\lambda\sigma} + \dfrac{q_\lambda q_\sigma}{M_1^2}}{q^2 + M_1^2} - \frac{q\ q_\sigma}{M_1^2(q^2 + M_2^2)} \end{gathered} \tag{83}$$

so $\tilde{\Delta}_{\lambda\sigma}(q) \sim 1/q^2$ for large q. The problem of course is that the interaction

$$\mathscr{L} = g \mathscr{J}_\lambda \tilde{\phi}_\lambda = g \mathscr{J}_\lambda \left(a_\lambda + \frac{i}{M_1} \partial_\lambda b \right) \neq \mathscr{L}^+ = g \mathscr{J}_\lambda \left(a_\lambda - \frac{i}{M_1} \partial_\lambda b \right). \tag{84}$$

Lee and *Wick* have tried to show, in a recent series of papers [8], that despite the non-Hermiticity of the Hamiltonian, one may introduce an indefinite metric $\eta = (-1)^{n_b}$ where n_b is the number of b particles in a state, and still have a unitary S-matrix. We do not have the time to go into these very interesting arguments, but recommend that the reader consult the articles in question.

IV. Conclusion

We have examined a series of models of weak interactions. For small momentum transfer, they all agree with the simple phenomenological interaction (1), but differ from it when calculations are performed to higher order in G_F, when it becomes necessary to know the behaviour for all momentum transfers. Since all the models give the same low energy phenomenological results, the choice is largely an aesthetic one as to which we prefer. The renormalizeable models, though free of the increasing divergence, seem too complicated to be really attractive and furthermore it seems unattractive that the $V-A$ nature of weak interactions should turn out to be an accident.

The unconventional non-renormalizeable models do not seem sompelling either. If there really is a cutoff, what is the cause of it? We have already mentioned strong interactions and weak interactions. *Ioffe* [30] has pointed out that perhaps electrodynamics provides the cutoff;

the electromagnetic interactions of the W are divergent and, with a cutoff Λ, behave like $(\alpha \Lambda^2/M_w^2)^m$.

He suggests that this will perhaps provide a natural cutoff at $\Lambda \sim M_w/\sqrt{\alpha}$. This does not seem to allow, however, for a cutoff $\Lambda \sim 3$–4 BeV as we found in the $K_L - K$; mass difference in a conventional theory as then we would need $M_w \sim 0.5$ BeV.

There is of course the mystery of how the W manages, if it exists, to have such a large mass. But then why does the muon have a mass two hundred ten times larger than the electron. I think it safe to say that a deep understanding of the weak interactions has not yet been reached. Perhaps new experiments will provide us with the necessary clues by finding significant derivations from our present old faithful current × current interaction.

References

1. For general notation see e.g. *Adler, S. L., Dashen, R. F.*: Current algebras. New York: W. A. Benjamin, Inc., 1968.
2. *Cabibbo, N.*: Phys. Rev. Letters **10**, 531 (1963).
3. *Kummer, W., Segrè, G.*: Nucl. Phys. **64**, 585 (1965).
4. *Christ, N.*: Phys. Rev. **176**, 2086 (1968).
5. *Shabalin, E. P.*: Yadern. Fiz. **8**, 74 (1968); — Soviet. J. Nucl. Phys. **8**, 74 (1969). *Galaktianov, Y., Shabalin, E. P.*: JETP Letters **8**, 369 (1968).
6. *Tahikawa, Y., Nakamura, S.*: Progr. Theoret. Phys. (Kyoto) Supp. **37–38**, 306 (1966).
7. *Fronsdal, C.*: Phys. Rev. **176**, 1846 (1968).
8. *Lee, T. D.*: Proceedings of Conference on Weak Interactions, held at CERN on Jan. 14–17, to be published.
   — *Wick, G. C.*: Nucl. Phys. Bg, 209 (1969).
9. *Mohapatra, R. N., Subba Rao, J., Marshak, R. E.*: Phys. Rev. Letters **20**, 1081 (1968).
10. *Low, F. E.*: Comments Nucl. Particle Phys. **2**, 33 (1968).
11. *Ioffe, B. L., Shabalin, E. P.*: Soviet. J. Nucl. Phys. **6**, 828 (1967).
12. *Feinberg, G.*: Phys. Rev. **134**, B 1255 (1964).
13. *Okubo, S.*: Nuovo Cimento A **54**, 491 (1968).
14. *Pepper, S. V., Ryan, C., Okubo, S., Marshak, R. E.*: Phys. Rev. B **137**, 1259 (1965).
15. *Ericson, T., Glashow, S.*: Phys. Rev. **133**, B 130 (1964).
16. *Callan, C. G.*: Phys. Rev. Letters **20**, 809 (1968).
17. *Dashen, R., Frautschi, S., Gell-Mann, M., Hara, Y.*: In: *Gell-Mann, M., Ne'eman, Y.*: The eightfold way, p. 254. New York: W. A. Benjamin 1964.
18. *D'Espagnat, B.*: Phys. Letters **7**, 209 (1963).
19. — *Villachon, Y.*: Nuovo Cimento **33**, 348 (1964).
20. *Ryan, C., Okubo, S., Marshak, R. E.*: Nuovo Cimento **34**, 753 (1964).
21. *Okubo, S.*: Phys. Letters **8**, 362 (1964).
22. *Good, M. L., Michel, L., de Rafael, E.*: Phys. Rev. **151**, 1194 (1966).
23. *McKellar, B. H. J.*: Phys. Rev. **178**, 2160 (1969).
24. *Segrè, G.*: Phys. Rev. **181**, 1996 (1969).
25. See e.g. *Bernardini, G.*: Lectures at 1964 «Scuola Internezionale di Fisica» held at Varenna (New York: Academic Press 1966).
26. *Gell-Mann, M.*: Physics **1**, 63 (1964).

27. *Li, L. F., Segrè, G.:* Higher order weak interactions in a model, to be published in Phys. Rev.
28. See for instance the calculation by *Glashow, S., Weinberg, S., Schnitzer, H.:* Phys. Rev. Letters **19**, 203 (1967), who find $M_w \sim 8$ Bev, neglecting divergent terms.
29. *Gell-Mann, M., Goldberger, M., Low, F., Kroll, N.:* Phys. Rev. **179**, 1518 (1969).
30. *Ioffe, B. L.:* Soviet Phys. JETP (English Transl.) **20**, 654, 1965.

Prof. Dr. *Gino Segrè*
University of Pennsylvania
Department of Physics
Philadelphia, Pa. 19104/U.S.A.

Weak Interactions at Small Distances*

H. Pietschmann

Contents

Abstract 193
1. Introduction 193
2. Nonlocal Effects in Weak Interactions 194
3. Digression on Higher Orders 196
4. Semi-leptonic Processes 196
5. μ-decay 198
6. Conclusion 199
References 200

Abstract

An S-operator for Weak Interactions is suggested which includes possible non-local structure effects. Leptonic and semi-leptonic processes are analyzed with this S-operator and a parameterization of the electron spectrum in μ-decay is suggested which allows for an experimental detection of such effects.

1. Introduction

Divergences in higher order Weak Interactions present the most persistent and difficult problem in this field. Though they do not immediately inflict upon the predictive power of the theory, they are the stumbling block which brings down any attempt to formulate a consistent theory of weak interactions. Over the decade, many bright ideas have been tried out [1], but it is fair to say that all attempts have failed so far.

In this paper, we leave the hard task of higher order Weak Interactions and look into the related problem of possible structure at small distances. It seems that attempts at the solution of the former problem are at least ambiguous unless the latter is clarified. Conventionally, the question as to whether there is structure in Weak Interactions is asked in the following way: "Is there an Intermediate Boson or not?" This

* Supported by „Österreichischer Forschungsrat".

question, however, is too narrow-minded, for the Intermediate Vector Boson Model (IVB-Model) is just one particular model which may or may not be the actual cause for structure in Weak Interactions. Therefore, the way we ask the question in this paper is the following: "Is there a structure in Weak Interactions at small distances?" Without reference to any specific model we want to analyze this question purely from empirical knowledge.

Before we enter into the actual analysis, a step aside may be allowed which shows that there *must* be some structure in every interaction, for a "point" is no physical object. This is seen from a gedanken-experiment much like the Heisenberg argument which led to the uncertainty relation. If distances are probed in space, photons are needed as an observational device. (Any other suitable particle will be just as good, of course.) The smaller the distance in question becomes, the shorter the wavelength of the photon has to be. At a certain critical length – Planck's length – the photon will be so heavy, that its own gravitational field will disturb the geometry of space so much that a separation of the endpoints is no longer possible. This is an absolute lower limit for physical distances [2]. Numerically, Planck's length is given by

$$\lambda_p = \sqrt{\frac{\hbar\kappa}{c}} \simeq 6 \cdot 10^{-33}\,\mathrm{cm} \tag{1}$$

where κ is the gravitational constant. This distance is certainly far beyond the structure we can presently probe with Weak Interaction experiments but it is worth while to keep the argument at the back of one's mind.

2. Nonlocal Effects in Weak Interactions

The first investigation of general structure effects in Weak Interactions goes back to *Lee* and *Yang* [3] in the "pre-V−A" period. Today, the theory of weak interactions is conventionally formulated by writing down a Hamiltonian [4]

$$H_W = \frac{-G}{\sqrt{2}} J_\lambda^+(x)\, J^\lambda(x)\,. \tag{2}$$

J_λ is the total current

$$J_\lambda(x) = l_\lambda(x) + \cos\theta\, j_\lambda(x) + \sin\theta\, \tilde{J}_\lambda(x) \tag{3}$$

where θ is the Cabibbo angle and j_λ and $\tilde{J}_\lambda$ are the currents transforming as π and K under SU_3. The leptonic current l_λ is

$$l_\lambda(x) = \sum_l \bar{\psi}_l(x)\, \gamma_\lambda(1+\gamma_5)\, \psi_{\nu_l}(x), \qquad l = \mathrm{e}, \mu\,. \tag{4}$$

In order to avoid the divergent higher orders, one usually requests as a supplementary condition that the Hamiltonian is used in lowest order only. Here, we like to make this explicit and hence we directly start with the S-operator instead of the Hamiltonian.

$$S = I - \frac{iG}{\sqrt{2}} \int d^4x \frac{1}{2} \{J^\lambda(x)\, l_\lambda^+(x) + \text{h.c.}\} + O(G^2). \tag{5}$$

Being fancy, we left out the non-leptonic part of the interaction so that we restrict ourselves to well-understood territory. Therefore we can state that the S-operator (5) is very well supported by experiments at the present range of energy and size. At low energies, the size is, of course, determined by the accuracy of experiments. Presently, we are down to roughly 10^{-13} cm. Nothing is known about $O(G^2)$, unfortunately.

Let us now carefully extend the S-operator (5) to include structure effects, without destroying its characteristic features which guarantee the good agreement with experiments. First of all, we stick to local action of the currents to lowest order in G. Secondly, the currents are taken in their conventional form without Fierz-reordering [5]. With these restrictions we arrive at

$$S = I - \frac{iG}{\sqrt{2}} \int d^4x\, d^4y \frac{1}{2} \{J^\lambda(x)\, K_{\lambda\nu}(x-y)\, l^{\nu+}(y) + \text{h.c.}\} + O(G^2) \tag{6}$$

where $K_{\lambda\nu}$ is an arbitrary structure function which has to be determined by experiment. Its Lorentz-decomposition is given by

$$K_{\lambda\nu}(x) = M^4 \left\{ g_{\lambda\nu} F_1(x^2) + \frac{1}{M^2} \frac{\partial^2}{\partial x^\lambda \partial x^\nu} F_2(x^2) \right\}. \tag{7}$$

$F_i(x^2)$ are invariant, dimensionless structure functions. M is some characteristic parameter determining the size of non-local effects. In general, it need not be related to an actual mass of an observable particle.

In order to keep structure functions dimensionless also in momentum space, we define the Fourier transform by

$$F_i(x^2) = (2\pi M)^{-4} \int d^4k\, e^{ikx} F_i(k^2), \qquad i = 1, 2. \tag{8}$$

In the IVB-Model, the structure functions are given by

$$\text{IVB:} \quad F_1(k^2) = F_2(k^2) = [(k^2/M^2) - 1 + i\varepsilon]^{-1}. \tag{9}$$

In this particular case, the parameter M becomes the mass of the intermediate boson and we have

$$\text{IVB:} \quad M = m_W. \tag{10}$$

Reiterating, we emphasize again that this is only a special case of the general investigation.

3. Digression on Higher Orders

Before we enter into an analysis of the S-operator (6), let us contemplate on what can be said about the higher orders $O(G^2)$. To do this, we go one step ahead and write the S-operator

$$S = I + iH + R + O(G^2). \tag{11}$$

Unitarity then tells us that

$$R + R^{+} + H^2 = 0. \tag{12}$$

Another condition stems from causality. If we invoke it in the form of Bogolubov, we write

$$H = -\frac{1}{\sqrt{2}} \int d^4x\, G(x)\, H(x). \tag{13}$$

Bogolubov-causality then requests

$$\frac{\delta^2 R(G)}{\delta G(y)\, \delta G(x)} + \frac{1}{2} H(y)\, H(x) = 0 \quad \text{for} \quad x \lesssim y. \tag{14}$$

It is immediately clear that even without reference to Lagrangian field theory, higher orders are largely influenced by the precise form of the first order. Research in this direction has been carried out by *Eoll* [6].

4. Semi-leptonic Processes

In this section, the influence of the structure functions on semi-leptonic processes shall be investigated.

A: $P \to l \nu_l$.

P represents a pseudoscalar meson, i.e. π or K. The matrix element for this decay is ordinarily characterized by a constant f_p. The S-operator (6) introduces the following change:

$$f_p \to f_p \left\{ F_1(m_p^2) - \frac{m_p^2}{M^2} F_2(m_p^2) \right\}. \tag{15}$$

It is seen, that the curly bracket cancels out in the $\mu - e$ ratio. This is clear since we kept local action of the lepton current.

B: $P \to \pi l \nu_l$.

For the 3 particle meson decay we use the standard form for the matrix element of the hadronic current [7]

$$\langle \pi(p_2) | J_\lambda(0) | P(p_1) \rangle \propto f_+(q^2)\,(p_1 + p_2)_\lambda - f_-(q^2)\, q_\lambda \tag{16}$$

with

$$q = p_2 - p_1 \,. \tag{17}$$

Taking into account the structure functions changes the form factors in the following way:

$$f_+(q^2) \to f_+(q^2)\, F_1(q^2)\,, \tag{18a}$$

$$f_-(q^2) \to f_-(q^2)\left[F_1(q^2) - \frac{q^2}{M^2} F_2(q^2)\right] - \frac{m_p^2 - \mu^2}{M^2} f_+(q^2)\, F_2(q^2) \tag{18b}$$

where μ is the pion mass.

C: $B \to B' l \nu_l$

B and B' represent two baryons. The baryonic matrix element is decomposed into the 6 standard form factors [7]

$$\langle B'(p_2)| J_\lambda(0) | B(p_1)\rangle \propto \bar{u}(p_2)\, \Gamma_\lambda(p_1, p_2)\, u(p_1) \tag{19}$$

with the vertex function Γ_λ given by

$$\begin{aligned} \Gamma_\lambda = {} & f_1(q^2)\, \gamma_\lambda + f_2(q^2)\, i\sigma_{\lambda\nu} q^\nu + f_3(q^2)\, q_\lambda \\ & + g_1(q^2)\, \gamma_\lambda \gamma_5 + g_2(q^2)\, q_\lambda \gamma_5 + g_3(q^2)\, i\sigma_{\lambda\nu} q^\nu \gamma_5 \,. \end{aligned} \tag{20}$$

The momentum transfer q is defined as in Eq. (17). Nonlocal effects cause the following change in the form factors:

$$f_i(q^2) \to f_i(q^2)\, F_1(q^2)\,, \qquad i = 1, 2\,, \tag{21a}$$

$$f_3(q^2) \to f_3(q^2)\left[F_1(q^2) - \frac{q^2}{M^2} F_2(q^2)\right] + \frac{m_B - m_{B'}}{M^2} f_1(q^2)\, F_2(q^2)\,, \tag{21b}$$

$$g_i(q^2) \to g_i(q^2)\, F_1(q^2)\,, \qquad i = 1, 3\,, \tag{21c}$$

$$g_2(q^2) \to g_2(q^2)\left[F_1(q^2) - \frac{q^2}{M^2} F_2(q^2)\right] - \frac{m_B + m_{B'}}{M^2} g_1(q^2)\, F_2(q^2) \tag{21d}$$

Note that the structure function $F_2(q^2)$ drops out in the limit of vanishing lepton mass [7].

It is fairly obvious that a detection of non-local effects in semi-leptonic processes is beyond the limits of present day Weak Interaction physics. Let us therefore turn to the purely leptonic μ-decay.

5. μ-decay

The electron energy spectrum in μ-decay is calculated from the V-A-theory, i.e. the S-operator (5) to be

$$N(x)=\frac{G^2 m_\mu^5}{96\pi^3}x^2(3-2x)\,. \tag{22}$$

In deriving Eq. (22), the electron mass has been neglected. x is the dimensionless electron energy varying in the interval [0, 1]

$$x=2E_e/m_\mu\,. \tag{23}$$

Normally, the energy spectrum (22) is then parameterized in the following way

$$N(x)=\frac{G^2 m_\mu^5}{16\pi^3}x^2\left[1-x+\frac{2}{3}\varrho\left(\frac{4}{3}x-1\right)\right] \tag{24}$$

where the Michèl-parameter ϱ is predicted by the V-A-theory [8] to be 3/4, as can be inferred from a comparison with Eq. (22). Tensor coupling would give $\varrho=1$ whereas scalar and pseudoscalar predict $\varrho=0$. The best experimental value at present is [9]

$$\varrho_{\mathrm{exp}}=0.752\pm0.003\,. \tag{25}$$

If we start from the S-operator with structure functions, Eq. (6), not all phase space integrations can be carried out in the most general case. As in the case of semi-leptonic processes, the structure function F_2 drops out in the limit of vanishing electron mass. The electron spectrum can be represented as an integral.

$$N(x)=\frac{G^2 m_\mu^5}{16\pi^3}\int_0^x d\sigma|F_1(\sigma)|^2(x-\sigma)(1-x+\sigma) \tag{26}$$

where σ is the dimensionless momentum transfer squared

$$\sigma=q^2/m_\mu^2\,. \tag{27}$$

Since x is always less than unity, only a very small part of $F_1(\sigma)$ enters the integral and an expansion is fully justified

$$F_1(\sigma)=1+\zeta\sigma+O(\sigma^2)\,. \tag{28}$$

ζ is the characteristic parameter for structure effects in μ-decay. In the special case of the IVB-model one has

$$\text{IVB:}\quad \zeta=m_\mu^2/M_W^2\,. \tag{29}$$

The electron spectrum (26) now becomes

$$N(x)=\frac{G^2 m_\mu^5}{96\pi^3}x^2\{3-2x+\zeta x(2-x)\}. \tag{30}$$

It is seen, that the spectrum can *no longer* be represented by a single Michèl-parameter. We suggest, that the parameterization (30) is used instead of (24) for a fit to experimental data. This presupposes accepting the V-A-theory [8]. However, in our days it seems that a search for structure effects is more important as well as more relevant than a further check of V-A. The way, both tests are combined by means of the formula [10]

$$\varrho=\frac{3}{4}+\frac{1}{3}\left(\frac{m_\mu}{m_W}\right)^2 \tag{31}$$

is a rather poor compromise in the days of precision experiments. If one insists in testing both V-A and structure effects, he can use the parameterization [3] of the *normalized* electron spectrum

$$n(x)=x^2\{6-4x+ag_1(x)+bg_2(x)\} \tag{32}$$

where

$$g_1(x)=\frac{1}{3}(3-4x), \tag{33a}$$

$$g_2(x)=\frac{1}{3}(3-8x+5x^2). \tag{33b}$$

Eq. (30) then requires

$$\text{V-A:} \quad a=2b=-\frac{12}{5}\zeta \tag{34}$$

whereas an S-P-theory would give

$$\text{S-P:} \quad a=3b=-\frac{6}{5}\zeta. \tag{35}$$

This parameterization therefore allows for both tests. Since the energy spectrum depends rather insensitively on the parameter b, we still suggest to use Eq. (30) for fits to the data.

6. Conclusion

We have analyzed leptonic and semi-leptonic weak processes in the light of the S-operator (6) which includes general structure functions. Our conclusion is that semi-leptonic decays do not prove to be fruitful for a test of non-local effects. On the other hand μ-decay provides a wonderful playground for this kind of investigation provided Eq. (30) is used to fit experimental data.

Unfortunately, we cannot conclude this paper without a grain of salt. Radiative corrections have not been mentioned at all, but it is well-known that they are large. In the case of the IVB-model, they can, of course, be calculated and subtracted from measured data. In the most general case, they are unknown since we do not know the electromagnetic properties of some obscure structure at small distances unless this structure is attributed to a special physical effect. It will need some investigation of special models to get an idea about radiative corrections in the general case. Right now, we suggest to use radiative corrections of the IVB-model and identify ζ by means of Eq. (29).

References

1. An incomplete list of typical examples is: *Feinberg, G., Pais, A.:* Phys. Rev. **131**, 2724 (1963); **133**, B 477 (1964).
   Pietschmann, H., Nilsson, J.: Phys. Rev. **142**, 1173 (1966).
   Gell-Mann, M., Goldberger, M., Kroll, N., Low, F.: Phys. Rev. **179**, 1518 (1969).
   Lee, T. D., Wick, G. C.: Nucl. Phys. B **9**, 209 (1969); B **10**, 1 (1969).
2. I am indepted to Prof. *R. Sexl* for bringing this point to my attention.
3. *Lee, T. D., Yang, C. N.:* Phys. Rev. **108**, 1611 (1957).
4. The notation follows: *Bjorken, J., Drell, S.:* Relativistic quantum fields. New York: McGraw-Hill 1965; except for a single change of sign in γ_5.
5. Theories with Fierz reordered currents are presently studied by *K. Draxler* and *H. Rupertsberger.*
6. *Eoll, C. K.:* Can. J. Phys. **47**, 135 and 153 (1969).
7. e.g. *Pietschmann, H.:* Acta Phys. Austr. Suppl. V, 88 (1968).
8. In the case of μ-decay, the two component neutrino assumption is sufficient to derive this prediction. I am indepted to Prof. *V. Telegdi* for reminding me to point this out.
9. *Rubbia, C.:* EPS Conference "The growth points in physics" at Florence, April 1969.
10. Cf. reference 3 as well as
    Oneda, S., Pati, J. C.: Phys. Rev. Letters **2**, 125 (1959).
    Lee, T. D., Yang, C. N.: Phys. Rev. 119, 1410 (1960).

Prof. Dr. *Herbert Pietschmann*
Institut für Theoretische Physik
A-1090 Wien/Austria, Boltzmanngasse 5

Physical Symmetries in the Framework of Quantum Field Theory

JAN T. ŁOPUSZAŃSKI

Contents

1. Introductory Remarks 201
2. Assumptions 203
3. Main Results 204
4. Discussion of the Results 208
References 213

1. Introductory Remarks

In many papers concerned with the symmetry problems in quantum field theory, e.g. *Goldstone*'s theorem and miscellanuous topics related to it, an algebra of quasilocal observables as well as a conserved (pseudo) vector current local with respect to the elements of the algebra and to itself are under investigation (see e.g. [1, 2]). It is well known that the necessary condition to get a properly defined "charge" which in turn gives rise to a one parameter group of the theory under consideration, is the local conservation of the current. If the energy-momentum spectrum has a mass gap then it is also a sufficient condition, no – so called – "Goldstone particles". The theorems proved up to now state the existence of the "charge". They do not, however, exclude the case that the "charge" vanishes. Of course, such a vanishing "charge" would not give rise to a symmetry.

It is of some interest to learn are we able to construct a locally conserved current of a scalar, local, irreducible quantum field in such a way that the current is local with respect to the field, transforms under the same Poincaré group representation as the field and gives rise to a conserved nonvanishing "charge". Our conjecture is that this can not be done for one real field (one complex field admits a gauge transformation only) provided there is a discrete one particle mass different from zero in the energy-momentum spectrum. Loosely speaking this would mean that in a theory of one scalar field with one massive particle one can not expect to get any one parameter symmetry group induced by a conserved (pseudo) vector current except the gauge group. The only symmetry groups we may find are the global symmetries generated by self-adjoint operators invariant under the Poincaré group transformation.

In this talk we are going to exhibit a more general problem of exploring physical symmetries in a theory of several scalar, real fields subjected to some fairly weak restrictions (see below [3]). Every set of l real and m complex scalar fields can always be expressed in terms of $l+2m$ real scalar fields. By a physical symmetry we mean not only internal symmetries, preserving strict locality, but a possibly larger class of symmetries mapping localized elements of the field algebra into localized elements almost local with respect to the former [4]. We show that every symmetry induced by a locally conserved (pseudo) vector current belongs to this class (see [5, 6]) and to emphasize this point we restrict ourselves to physical symmetries which commute with the Poincaré group and with the *TCP* operator. We succeeded in giving the recipe how to find the most general form of such a symmetry, provided it is unitarily implemented in the Hilbert space. The essential thing is that the transformed incoming and outgoing fields are expressed as linear combination of the original ones, even if we do not demand a common *TCP* operator for the transformed and untransformed fields. Incidentally we showed also that the symmetry groups are compact.

A linear rearrangement of the original fields, suitable for physical problems, yields then a possibility of a classification of the relevant symmetries occurring in the theory of n scalar – not necessarily real – fields. In this way we get a partial answer to the question which of the pair- and self-conjugate multiplets, e.g. of the $SU(2)$ or $SU(3)$ groups, may appear in the theory [7].

For the $SU(2)$ symmetry we get the generalized Carruthers Theorem [8] which reads: self-conjugate isospin multiplets of bosons with integral isospin exist (odd number of fields), those with half-integral isospin, however, do not (even number of fields). For an even number of fields we may construct a unitary reducible representation of $SU(2)$ yielding two pair-conjugate multiplets.

In case of $SU(3)$ only the representations of 1, 8, 27 etc. dimensions are proper candidates for self-conjugate multiplets. The representations of other dimensions as 3, 6, 10 etc. may appear as pair-conjugate multiplets in a theory of 6, 12, 20 etc. real fields resp.

The results mentioned before are all derived for the theory of several scalar fields only. The formalism can be, however, extended to fields of a different transformation character with respect to the Poincaré transformation as bispinors or vectors. For the case of bispinor fields we carried out the computations in detail, but we are not going to reproduce them here. It turns out that the conclusions drawn for the physical symmetries are the same as in case of scalar fields irrespectively of the transformation character of the fields in the Minkowski space.

If we add to our basic assumptions a physically justified rule that *the massive bispinor fields are always complex*, then in case of the $SU(2)$ symmetry we can not expect for them any self-conjugate multiplets. We may, however, expect pair-conjugate multiplets of every (integer) dimension. For the case of $SU(3)$ representations of 1, 3, 6, 8, 10, etc. dimensions are admissible only as pair-conjugate multiplets.

2. Assumptions

To start with we list the *relevant* assumptions:

i) The Poincaré group is unitarily implemented in the Hilbert space $\mathscr{H}$. We denote the representation of (Λ, a) by $U(\Lambda, a)$. The spectrum of P^2, where

$$T(a) = U(1, a) = \exp\{iPa\}$$

consists of two discrete points $\mu^2 = 0$ (corresponding to the single vacuum state Ω) and $\mu^2 = m^2 \neq 0$ (corresponding to one particle states) as well as out of a continuous part, say $(4m^2, \infty)$ (scattering states).

ii) We are concerned with n scalar, real quantum fields $\psi_j(x) j = 1, \ldots, n$. All of them transform under the same Poincaré representation, viz.

$$U(\Lambda, a)\, \psi_j(x)\, U(\Lambda, a)^+ = \psi_j(\Lambda x + a)\,; \quad j = 1, \ldots, n \tag{1}$$

are local and local with respect to each other*. We assume for simplicity reasons that

$$(\Omega, \psi_j(f)\, \Omega) = 0, \quad j = 1, \ldots, n \tag{2}$$

where

$$\psi_j(f) \equiv \int \psi_j(x)\, f(x)\, \mathrm{d}x\,, \quad f(x) \in \mathscr{S}_4\,.$$

iii) To exclude trivial cases (as e.g. fields like $\psi_3 = a\psi_1 + b\psi_2$) and in order to get proper asymptotic conditions [9] we assume that

$$\sum_{j=1}^{n} a_j E_1\, \psi_j(f)\, \Omega = 0 \quad \text{if} \quad a_j = 0\,. \tag{3}$$

Here E_1 is the projection operator on the Hilbert subspace $\mathscr{H}_1$ labelled by $p^2 = m^2, p_0 > 0$ (spanned by the one particle states induced by the fields $\psi_j(x)$).

iv) Then the free asymptotic fields exist and we assume then to be irreducible. This implies that the original set of n fields $\psi_j(x)$ is also

* We believe that weak locality would be sufficient for our aims.

irreducible. As a matter of fact it would be sufficient to assume only that the incoming fields form an irreducible set (due to the existence of a common TCP operator).

We may form linear combinations out of these fields. They will be again local, local with respect to each other, real, will transform under the same representation of the Poincaré group as ψ_j and build an irreducible set. Let us call them $\varphi_j(x)\, j=1, \ldots, n$. We demand in addition that

$$(\Omega, \varphi_j(x)\, E_1\, \varphi_k(y)\, \Omega) = i\, \delta_{jk}\, \Delta^{(+)}(x-y\,; m^2)\,, \qquad j, k = 1, \ldots, n\,. \quad (4)^*$$

These fields have asymptotic free fields

$$\varphi_{j, \substack{\text{in} \\ \text{out}}}(x) \equiv \varphi_{j, \text{ex}}(x)$$

(in virtue of (3)) belonging all to the same mass $m^2 \neq 0$. In the sequel we shall take the fields $\varphi_j(x)$ as basic.

3. Main Results

Our main Statement reads as follows:

Statement

Let us assume that there exists such a unitary transformation V for which

$$(\Omega, V\Omega) \neq 0 \tag{5}$$

and the fields

$$V\, \varphi_j(x)\, V^+ = A_j(x) \tag{6}$$

transform under the same representation of the Poincaré group and under the same TCP operator as $\varphi_j(x)$, viz.

$$U(\Lambda, a)\, A_j(x)\, U(\Lambda, a)^+ = A_j(\Lambda x + a)\,; \qquad j = 1, \ldots, n \tag{7a}$$

$$\Theta\, A_j(x)\, \Theta^{-1} = A_j(-x) \tag{7b}$$

and are almost local with respect to $\varphi_k(x)\, k = 1, \ldots, n$, i.e.

$$R^N \left\langle \prod_{j=1}^{s} C_{l_j}(x_j) \right\rangle_T \xrightarrow[R \to \infty]{} 0 \tag{8}$$

where N is any real integer, the index "T" stands for "truncated" and C_{l_j} mean either A or φ. The variables x_j are clustered so that the 3-dimensional distance between some two clusters is R. Then we have

$$V\, \varphi_{j, \text{ex}}(x)\, V^+ = \sum_{k=1}^{n} \alpha_{jk}\, \varphi_{j, \text{ex}}(x) \tag{9}$$

---

* The case $(\Omega, \varphi_j(x)\, E_1\, \varphi_j(y)\, \Omega) = 0$ is excluded by assumption (3).

where α_{jk} form an n-dimensional (real) orthogonal matrix and

$$[V, S] = 0 \tag{10}$$

where S is the scattering matrix. V is uniquely defined by (9) up to a constant number of modulus 1.

Proof

From the fact that φ_j and A_k, $j, k = 1, \ldots, n$ transform under the same representation of the Poincaré group and from the irreducibility of φ_j follows

$$V^+ U(\Lambda, a)^+ V U(\Lambda, a) = \lambda 1 ; \qquad \lambda \text{ a number}$$

consequently

$$(\Omega, V\Omega) = \lambda(\Omega, V\Omega) .$$

Because of (5) we have $\lambda = 1$. Thus V commutes with the Poincaré group

$$[U(\Lambda, a), V] = 0 \tag{11}$$

and

$$U(\Lambda, a)\, V\Omega = V\Omega .$$

The uniqueness of the vacuum and the unitarity of V implies then

$$V\Omega = e^{ib}\Omega , \qquad b \text{ real} .$$

We incorporate the factor e^{ib} into V so that we get

$$V\Omega = \Omega . \tag{12}$$

Let V act on the one particle states

$$\varphi_{j,\mathrm{ex}}(f)\, \Omega \equiv \Phi_j(f) .$$

In virtue of (4) and the stability of the one particle states the vectors $\Phi_j(f)$ and $\Phi_k(g)$ for $j \neq k$ are orthogonal to each other. In virtue of (11) we have for the vector valued distribution

$$\Phi_j(p) \equiv \tilde{\varphi}^+_{j,\mathrm{ex}}(\boldsymbol{p}, \omega)\, \Omega \quad \text{with} \quad \omega = +\sqrt{\boldsymbol{p}^2 + m^2} ,$$

$$U(1, a)\, V\, \Phi_j(p) = V\, U(1, a)\, \Phi_j(p) = e^{ipa}\, V\, \Phi_j(p) .$$

The irreducibility of the fields $\varphi_{j,\mathrm{ex}}$ implies then

$$V\, \Phi_j(p) = \sum_{k=1}^{n} \alpha_{jk}(p)\, \Phi_k(p) . \tag{13a}$$

Applying to both sides of (13a) the operator $U(\Lambda, 0)$ and making use once more of (11) we get

$$V\, \Phi_j(\Lambda p) = \sum_{k=1}^{n} \alpha_{jk}(p)\, \Phi_k(\Lambda p) .$$

Hence α_{jk} does not depend on p, and we wind up with

$$V\,\Phi_j(f) = \sum_{k=1}^{n} \alpha_{jk}\,\Phi_k(f) \tag{13b}$$

where α_{jk} are independent on the choice of the test function f. Since $\Phi_j(f)$ and $\Phi_k(g)$ are orthogonal for $j \neq k$ we infer from (13) that

$$V^+\,\Phi_j(f) = \sum_{k=1}^{n} \overline{\alpha_{kj}}\,\Phi_k(f) \tag{14}$$

and since V commutes with the TCP operator

$$\alpha_{jk} = \overline{\alpha_{jk}} \tag{15a}$$

the numbers α_{jk} are real. Moreover, V is unitary therefore

$$\sum_{j=1}^{n} \alpha_{jk}\alpha_{jl} = \delta_{kl}\,. \tag{15b}$$

The matrix α_{jk} is an n-dimensional (real) orthogonal matrix.

In order to get further conclusions for the fields themselves let us use the generalized procedure of *Borchers* [10]. The original result of Borchers said that for every two scalar real fields A and B, local with respect to each other and to itself, one of the fields being irreducible, the following is true

$$[A_{\mathrm{ex}}(x), B_{\mathrm{ex}}(y)] = i\,\Delta(x-y\,; m^2)$$

where

$$\Delta(x\,; m^2) = -\frac{i}{(2\pi)^3}\int \mathrm{d}p\, \mathrm{e}^{-ix\cdot p}\,\varepsilon(p)\,\delta(p^2-m^2) \tag{16}$$

or

$$A_{\mathrm{ex}}(x) = \pm B_{\mathrm{ex}}(x)$$

where A_{ex} and B_{ex} are asymptotic free fields belonging to the same mass $m \neq 0$. The proof of Borchers is based on an idea of Zimmermann and uses the L.S.Z. formalism. The generalized procedure uses more advanced mathematical tools. We are not going to enter into details here and confine ourselves to state the result only. In our case, under the assumptions listed before, we get

$$[\varphi_{j,\mathrm{ex}}(x), A_{k,\mathrm{ex}}(y)] = \text{a number valued distribution}\,. \tag{17}$$

Since the mapping induced by V is a global one the relation (6) can be immediately extended to the asymptotic limits, viz.

$$A_{k,\mathrm{ex}}(x) = V\,\varphi_{k,\mathrm{ex}}(x)\,V^+\,. \tag{18}$$

In order to evaluate (17) we make use of (12), (13), (14), (15)* and (18). The irreducibility of the fields $\varphi_{j,\text{ex}}$ as well as (12), (17), and (18) lead us to the conclusion that

$$A_{k,\text{ex}}(x) \equiv V\, \varphi_{k,\text{ex}}(x)\, V^{+} = \sum_{j=1}^{n} \alpha_{kj}\, \varphi_{j,\text{ex}}(x)\,. \tag{9}$$

Notice that V is uniquely defined by (9) due to the irreducibily of the fields $\varphi_{k,\text{ex}}$ as well as because of (12)**.

Since (9) holds for the incoming as well as for the outgoing fields V satisfies also (10).

This accomplishes the proof.

It is worthwhile to emphasize that the assumption that the sets of fields A_k and φ_j are almost – or quasi-local with respect to each other is not sufficient to obtain Borchers result (9). What one gets instead is

$$V\, \varphi_{k,\text{ex}}^{(-)}\, V^{+} = \sum_{j=1}^{n} \alpha_{kj}\, \varphi_{j,\text{ex}}^{(-)}$$

where $\varphi^{(-)}$ denotes the negative frequency part of φ. α_{kj} do not need to be real (see [10]), although the matrix remains still unitary. This difficulty can be overcome most simply by assuming (7b)***. It is, however, possible that it can be overcome in a less drastic way. Notice also that in order to get (15a) we had to make use of (7b).

---

* To make things clear let us consider e.g.

$$\begin{aligned}&(\varphi_{j,\text{ex}}(x)\,\Omega,\, A_{k,\text{ex}}(y)\,\Omega) - (A_{k,\text{ex}}(y)\,\Omega,\, \varphi_{j,\text{ex}}(x)\,\Omega)\\ &\quad= (\varphi_{j,\text{ex}}(x)\,\Omega,\, V\varphi_{k,\text{ex}}(y)\,\Omega) - (\varphi_{k,\text{ex}}(y)\,\Omega,\, V^{+}\,\varphi_{j,\text{ex}}(x)\,\Omega)\\ &\quad= i\alpha_{kj}\,\Delta(x-y;m^2)\end{aligned}$$

where $\Delta(x;m^2)$ is given by (16).

** Suppose on the contrary that there are two different solutions to (9), say V and V', then

$$V'^{+}\, V\varphi_{k,\text{ex}}(f)\, V^{+}V' = \varphi_{k,\text{ex}}(f) \quad \text{and} \quad V'^{+}\,V = \mu 1$$

a number. Due to (12) we have $\mu = 1$.

*** The author is obliged to Dr. *R. Haag* and Dr. *M. Rinke* for pointing out this difficulty and suggesting to him that a common *TCP* operator is sufficient to exclude cases like that of two real fields $\varphi_{\text{ex}} = \varphi_{\text{ex}}^{(+)} + \varphi_{\text{ex}}^{(-)}$ and $A_{\text{ex}} = e^{i\lambda}\,\varphi_{\text{ex}}^{(+)} + e^{-i\lambda}\,\varphi_{\text{ex}}^{(-)}$ which are quasilocal with respect to each other for each λ (λ is real), nevertheless are not equal up to a constant factor. As a matter of facts this particular case can be excluded in our case since

$$e^{-i\lambda N_{\text{ex}}}\,\varphi_{\text{ex}}^{(-)}(x)\, e^{i\lambda N_{\text{ex}}} = e^{-i\lambda}\,\varphi_{\text{ex}}^{(-)}(x)$$

where

$$N_{\text{ex}} = \tfrac{1}{2}\int dp\,\delta(p^2-m^2)\,\theta(p) : \tilde{\varphi}^{+}(p)\,\tilde{\varphi}(p):$$

and

$$A_{\text{ex}} = V\varphi_{\text{ex}}V^{+} = e^{-i\lambda N_{\text{ex}}}\,\varphi_{\text{ex}}\,e^{i\lambda N_{\text{ex}}}$$

which implies (irreducibility of φ_{ex} and stability of the vacuum under V)

$$V = \exp\{i\lambda N_{\text{in}}\} = \exp\{i\lambda N_{\text{out}}\}$$

what can not be true unless $N_{\text{in}} = N_{\text{out}}$ (only elastic scattering admissible). But as soon as more then one fields are considered the problem becomes more complicated.

4. Discussion of the Results

The most general unitary transformation satisfying the hypotheses of the Statement is given by (9), (10), and (15). The essential part of the Statement is that the transformed incoming or outgoing fields are expressed as linear combinations of the original fields. Even if we do not demand a common TCP operator for the transformed and nontransformed fields the relation invoqued by the unitary transformation maintains its linear character.

Thus we may exclude as nonphysical (in the sense of *Haag* and *Kastler*) all transformations which give rise to nonlinear expressions (e.g. the scattering matrix $S \neq 1$ for a free field φ_{in} is a nonphysical transformation; this proves at the same time that the fields φ_{in} and φ_{out} can never be almost – or quasi-local with respect to each other for a nontrivial theory). What was said so far does not mean that for every given orthogonal finite dimensional matrix one can construct a unitary operator V. This can be seen clearly on the example of the gauge transformation. It is obvious that every free complex scalar field is gauge invariant; this is in particular true for the complex free fields C_{in} and C_{out}; to each of these fields belongs a gauge generator, a charge Q_{ex}, induced by a conserved current $j_{ex}(x)$; but these two operators do not need to coincide; they are both Lorentz invariant quantities which, however, do not need to commute with $S \neq 1$, unless they coincide. Should we be able to prove the existence of a current, local with respect to the interpolating field $C(x)$ and giving rise to a non vanishing charge Q we were then in the position to state $Q = Q_{in} = Q_{out}$.

It is also worthwhile to emphasize once more that the assumption of the existence of a mass gap is of a paramount importance for getting our results.

From the Statement follows immediately the:

Corollary 1

Let us consider an abstract symmetry group G implemented unitarily by $V(g)$, $g \in G$, satisfying the restrictions mentioned above. The operators $V(g)$ transform n independent scalar, not necessarily real, fields which, however, are built out of n real fields φ_k. Should the symmetry be a physical one it has to have an n-dimensional representation of $SO(n)$. This implies in turn that in order to get physical symmetries for bosons we have only to look after such n-dimensional representations of the group G which can be made real by applying a suitable unitary transformation [7, 8].

Since the matrices $\alpha_{jk}(g)$ form an irreducible n-dimensional unitary representation of the group G we have the

Corollary 2

If the set of the unitary matrices $V(g)$ satisfying the hypotheses of the Statement forms a representation of a group $G(g \in G)$ this group is compact.

It is obvious that the internal symmetries fall into the class of unitary transformations obeying the requirements of our Statement. The same is also true for physical symmetries whose generators are induced by (pseudo) vectors (see [6]). To make this plausible consider e.g. a real generator

$$Q \equiv \int j_0(x)\, \mathrm{d}^3 x$$

where $j_\mu(x)$; $\mu = 1, 2, 3, 0$ is a (pseudo) vector, local with respect to the fields φ_k (it follows then that $j_\mu(x)$ is local itself), locally conserved, viz.

$$\frac{\partial j_\mu(x)}{\partial x_\mu} = 0$$

and transforming according to

$$U(\Lambda, a)\, j_\mu(x)\, U(\Lambda, a)^+ = \Lambda_\mu^{-1\nu} j_\nu(\Lambda x + a)\,.$$

Then it can be shown that Q is a Poincaré invariant quantity [2]. Since j_μ and φ_k have a common TCP operator and

$$\Theta\, j_\mu(x)\, \Theta^{-1} = -j_\mu(-x)$$

we have

$$\Theta\, Q\, \Theta = -Q\,;$$

then

$$V(\alpha) \equiv \exp\{i\alpha Q\}$$

satisfies (5) and (7).

It satisfies also (8); disregarding the domain questions let us consider the fields

$$A_l(x) \equiv i[Q, \varphi_l(x)]\,.$$

These fields are real, local with respect to φ_l and to themselves since j_μ is local with respect to φ_l; transform under the same representation of the Poincaré group as φ_l. We have also

$$E_1 A_l(f)\, \Omega = i E_1 Q\, \varphi_l(f)\, \Omega \neq 0$$

unless $QE_1 = 0$. In a manner similar to that out lined earlier we get

$$i[Q, \varphi_{l,\mathrm{ex}}(x)] = \sum_{k=1}^{n} \beta_{lk}\, \varphi_{k,\mathrm{ex}}(x)$$

with β_{lk} real and antisymmetrical. By iterating this procedure we get e.g.

$$-[Q[Q,\varphi_{l,\mathrm{ex}}(x)]]=\sum_{k=1}^{n}\beta_{lk}[Q,\varphi_{k,\mathrm{ex}}(x)]=\sum_{k=1}^{n}(\beta^2)_{lk}\varphi_{k,\mathrm{ex}}(x)$$

and finally

$$e^{i\alpha Q}\,\varphi_{l,\mathrm{ex}}(x)\,e^{-i\alpha Q}=\sum_{k=1}^{n}(e^{\alpha\beta})_{lk}\,\varphi_{k,\mathrm{ex}}(x)$$

where α is a real number.

However, for the real generators [6]

$$Q_\mu=\int T_{0\mu}(x)\,d^3x \tag{19}$$

induced by a tensor $T_{\nu\mu}(x)$; $\nu,\mu=0,1,2,3$ displaying similar properties to $j_\mu(x)$ mutatis mutandis the conditions imposed in the Statement (see especially condition (7)) lead to a trivial result. The same holds true for any generator set of tensorial character in the Minkowski space of rank $n\geqq 1$.

The most natural way to incorporate e.g. the case of the generators (19) into our scheme would be to change partly the premises of our Statement. Instead of (7) we should demand for

$$V(b)\equiv\exp\{ib_\mu Q^\mu\},\qquad b_\mu\text{ - real numbers}$$

the relations

$$U(\Lambda,a)\,V(b)\,U(\Lambda,a)^+=V(\Lambda b), \tag{20a}$$

$$\Theta\,V(b)\,\Theta^{-1}=V^+(b)=V(-b) \tag{20b}$$

to hold. Then it is easy to show that the transformation V is uniquely defined (up to a constant) by the equation

$$V(b)\,\varphi_{j,\mathrm{ex}}(x)\,V(b)^+=\sum_m\sum_l\beta_{jm}^{-1}\beta_{ml}\,\varphi_{l,\mathrm{ex}}(x+\lambda_m b) \tag{21}$$

where β_{jk} is a real, orthogonal matrix and λ_k are the eigenvalues of the matrix γ_{jk}, real and symmetrical, given by

$$Q^\nu\,\Phi_l(p)=\sum_{k=1}^{n}p^\nu\gamma_{lk}\,\Phi_k(p)\,.$$

Assuming that all n fields φ interact effectively among themselves it can be shown [6] that $\lambda_m=\lambda$, $m=1,\dots n$ and consequently

$$Q_\mu=\lambda P_\mu\,.$$

Returning to our previous considerations related to our Statement we find that for one real scalar field ($n=1$) we have

$$V\,\varphi_{\mathrm{ex}}V^{+}=\pm\varphi_{\mathrm{ex}}$$

since according to (15) $|\alpha|^2=1$ and $\alpha=\bar{\alpha}$. There is no continuous group of symmetry transformations different from a constant.

In case $n=2$ the 2-dimensional real representation of the one parameter group $SO\,(2)$ is reducible. By introducing the complex field

$$C(x)=\frac{1}{\sqrt{2}}(\varphi_1(x)+i\,\varphi_2(x)) \tag{22}$$

and its complex conjugate we may get

$$V(\alpha)\,C_{\mathrm{ex}}(x)\,V(\alpha)^{+}=e^{i\alpha}C_{\mathrm{ex}}(x) \qquad \alpha \text{ real number} \tag{23}$$

and $V(\alpha)$ is the gauge group transformation. For any 2-dimensional orthogonal matrix we have either (23) or

$$V C_{\mathrm{ex}}(x)\,V^{+}=e^{i\beta}C_{\mathrm{ex}}^{+}(x)\,, \qquad \beta \text{ real number} \tag{24}$$

where $V=V^{+}=V^{-1}$.

In case $n=3$ we may introduce the fields

$$C_1(x)=\frac{1}{\sqrt{2}}(\varphi_1(x)+i\,\varphi_2(x))=C_3^{+}(x)\,, \tag{25a}$$

$$C_2(x)=\varphi_3(x)\,, \tag{25b}$$

and consider the 3-dimensional complex irreducible representation of $SO\,(3)$ or $SU\,(2)$. In this way we get a triplet encompassing particles and its antiparticles as well, a self-conjugate triplet, like in case of $\pi^{+}, \pi^{0}, \pi^{-}$. There is, however, no nontrivial 2-dimensional group transforming C_1 and C_3 among themselves,

$$V C_{1,\mathrm{ex}}V^{+}=u C_{1,\mathrm{ex}}+v C_{3,\mathrm{ex}}\,; \qquad uv\neq 0 \text{ etc.}$$

leaving C_2 unchanged. Should C_2 remain unchanged uv has then to vanish.

In case $n=4$ we may introduce the fields

$$C_1(x)=\frac{1}{\sqrt{2}}(\varphi_1(x)+i\,\varphi_2(x))=C_3^{+}(x)\,, \tag{26a}$$

$$C_2(x)=\frac{1}{\sqrt{2}}(\varphi_3(x)+i\,\varphi_4(x))=C_4^{+}(x)\,. \tag{26b}$$

Then in addition to the 6-parameter 4-dimensional complex irreducible representation of $SO(4)$ we may also consider the 3-parameter 2-dimensional irreducible representation of $SU(2)$ and its complex conjugate. The letter ones will transform C_1, C_2 and C_3, C_4 resp. among themselves, viz.

$$\left.\begin{aligned} C_1' &= \alpha C_1 + \beta C_2 \\ C_2' &= -\bar{\beta} C_1 + \bar{\alpha} C_2 \end{aligned}\right\} |\alpha|^2 + |\beta|^2 = 1\,, \quad \alpha, \beta\text{-complex}\,. \tag{27}$$

Thus we get two pair-conjugate doublets, like in case of K^+ and K^0. However, there is no group transforming the fields C_1 and $C_3 = C_1^+$ among themselves. Other groups are trivial extensions of the cases considered earlier.

These particular results, especially that for $n=3$ and 4, can be extended to the case of an n-dimensional representation of $SO(n)$ for an arbitrary integer n and its subgroups [7].

We get is this way e.g. the so called generalized Carruthers Theorem [8] quoted in the Introductory Remarks of this talk. The proof follows immediately from the observation that each $n=(2m+1)$-dimensional representation of $SU(2)$, $m=0, 1, 2, \ldots$ can be made real (see e.g. case $n=3$ above), while the $n=2m$-dimensional (spinor) representation can not (see e.g. case $n=4$ [11]).

The same reasoning applied in case of $SU(3)$ rules out of considerations other self-conjugate multiplets as those of 1, 8, 27 etc. dimensions for bosons (see [11]).

If we have $2n$ real fields we can use a subgroup of $2n$-dimensional representation of $SO(2n)$ consisting of the matrices of the form

$$\begin{pmatrix} A, & B \\ -B, & A \end{pmatrix} \tag{28}$$

where A and B are real and $A+iB$ are unitary, n-dimensional matrices. Then one cast the matrices (28) unitary transformation

$$\frac{1}{\sqrt{2}} \begin{pmatrix} 1, & -i1 \\ -i1, & 1 \end{pmatrix}$$

into the form

$$\begin{pmatrix} A+iB, & 0 \\ 0\;, & A-iB \end{pmatrix} \tag{29}$$

The original fields are then transformed into two pair-conjugate multiplets. The inverse is also true, i.e. every matrix of type (29) can be brought into the shape (28). This insures that we may expect to have pair-conjugate multiplets of $SU(2)$ of n dimensions, $n=1, 2$, etc. and of $SU(3)$ of $n=1, 3, 6, 8, 10$, etc. for $2n$ fields.

Going over to the discussion of the case of bispinor fields we may assume the additional rule mentioned in the Introductory Remarks and claiming that the massive bispinor fields should be complex ones. It can be shown that the transformation character of the fields in the Minkowski space does not affect their behavior with respect to the physical symmetries [13]. In virtue of our additional restriction we have always to deal with an even number of *real* bispinor fields. Therefore we cannot expect to have any self-conjugate multiplets of $SU(2)$ for n *complex* fermion fields in the theory, as $2n$-dimensional representations of $SU(2)$ cannot be made real. We may however, always find n-dimensional pair-conjugate multiplets for n complex bi-spinor fields by applying the theorem concerning (28) and (29).

For the case of $SU(3)$ the representations of n dimensions where $n=(2m)^3$, $m=1,2,3$, could be regarded as good candidates for self-conjugate multiplets of $n/2$ complex fermion fields. However, the n dimensional space can be decomposed into a direct sum of subspaces invariant with respect to the $SU(2)$ transformations. Each representation of $SU(2)$ of odd number of dimensions appears an odd number of times. E.g. for $n=8$ we have the representations of $SU(2)$ of 3, 2, 2 and 1 dimensions. The odd number dimensional representations cannot be only pair-conjugate since they appear an odd number of times in the decomposition. Thus at least one of them has to be self-conjugate; this is, however, excluded by the Carruthers Theorem and our additional assumptions concerning fermion fields. Therefore we cannot expect to find self-conjugate multiplets for fermions.

Applying the theorems expressed by (28) and (29) we conclude that to every admissible n-dimensional representation of $SU(3)$ can correspond a pair-conjugate multiplet of n complex fermion fields.

Acknowledgment. The author is grateful to Mr. *A. Z. Jadczyk*, *B. Jancewicz* and Miss *L. Nikolova* as well as to Dr. *G. F. Dell'Antonio*, *J. Olszewski*, *H. Reeh*, and *K. Zalewski* for discussions.

References

1. *Kastler, D., Robinson, D. W., Swieca, A.:* Commun, Math. Phys. **2**, 108 (1966). — *Schroer, B., Stichel, J.:* Commun. Math. Phys. **3**, 258 (1966). — *Swieca, A.:* Nuovo Cimento (10) **52**, 242 (1967).
2. See review articles on symmetries: *Łopuszański, J.:* Fortschr. Physik **15**, 681 (1967). — *Reeh, H.:* Lecture Notes at the Vth Winter School of Theoretical Physics, Karpacz, 1968, Acta Univ. Wratisl. Vol. 99, Wrocław, 1968. — *Robinson, D. W.:* Lecture Notes at the NATO Summer School in Theor. Phys. Istanbul. 1966, CERN preprint 66 (1205) 5-Th. 708.
3. *Łopuszański, J. T.:* Commun. Math. Phys., in press.
4. *Haag, R., Kastler, D.:* J. Math. Phys. **5**, 848 (1964).
5. *Łopuszański, J. T.:* Commun. Math. Phys. **12**, 80 (1969); — Rep. Math. Phys., in press.

6. *Orzalesi, C. A., Sucher, J., Woo, C. H.:* Phys. Rev. Letters **21**, 1550 (1968).
7. *Jancewicz, B., Łopuszański, J. T., Nikolova, L.:* Submitted for publication in Nucl. Phys.
8. *Carruthers, P.:* Phys. Rev. Letters **18**, 353 (1967). — *Jin, Y. S.:* Phys. Letters B **24**, 411 (1967). — *Fleming, G. N., Kazes, E.:* Phys. Rev. Letters **18**, 764 (1967). — *Lee, H.:* Phys. Rev. Letters **18**, 1098 (1967). — *Kantor, P. B.:* Phys. Rev. Letters **19**, 394 (1967). — *Steinmann, O.:* Phys. Rev. Letters B **25**, 234 (1967). — *Fabri, E., Picasso, L. E.:* Nuovo Cimento, in press. — *Mathews, P. M., Jayaraman, J., Seethraman, M.:* Nuovo Cimento (10) A **58**, 625 (1968). — *Yee, D. D. H.:* Phys. Rev. **172**, 1537 (1968).
9. *Haag, R.:* Phys. Rev. **112**, 669 (1958). — *Hepp, K.:* Helv. Phys. Acta **37**, 639 (1964). — *Hepp, K.:* Commun. Math. Phys. **1**, 111 (1965). — *Araki, H., Haag, R.:* Commun. Math. Phys. **4**, 77 (1967).
10. The original version due to *Borchers, H. J.:* Nuovo Cimento (X) **15**, 784 (1960) was hereafter generalized by *H. Araki*; see *Hepp, K.:* Lecture Notes of the Brandais Summer School, Waltham, 1965 as well as *Araki, H., Haag, R., Schroer, B.:* Nuovo Cimento (10) **19**, 90 (1961); see, however, the remark concerning the role of the *TCP* invariance made in the talk (page 10).
11. *Wigner, E. P.:* Group theory, p. 288. New York: Academic Press 1959. – *Barcy. H.:* Leçons sur la theorie des groupes et les symmétries des particules elementaires. Paris: Gordon and Breach 1968.
12. *Helgason, S.:* Differential geometry and symmetric spaces. New York: Academic Press 1962.
13. A note on this subject, concerning the bispinor fields, will be published soon in a preprint form.

Prof. Dr. *Jan T. Łopuszański*
Institute of Theoretical Physics
University of Wrocław
Wrocław, Poland

GPSR Compliance

The European Union's (EU) General Product Safety Regulation (GPSR) is a set of rules that requires consumer products to be safe and our obligations to ensure this.

If you have any concerns about our products, you can contact us on ProductSafety@springernature.com

In case Publisher is established outside the EU, the EU authorized representative is:

Springer Nature Customer Service Center GmbH
Europaplatz 3
69115 Heidelberg, Germany

Batch number: 10620141

Printed by Printforce, the Netherlands